Free:
The End of the Human Condition

The Biological Reason Why Humans
Have Had To Be Individual,
Competitive, Egocentric and Aggressive

Jeremy Griffith

Published by
Foundation For Humanity's Adulthood
GPO Box 5095, Sydney NSW 2001, Australia
Phone: (02) 9486 3308

Website: **www.humancondition.info**

First published January 1988

Copyright © Jeremy Griffith 1988

Published by:
Centre For Humanity's Adulthood
Box 5095
G.P.O. Sydney, 2001
Australia

Written by Jeremy Griffith.
Funded by Gervase and Jeremy Griffith.
Assisted by Ann Williams.

Dedication and Acknowledgement:
To my Father and from my Mother.

ISBN 0 7316 0495 4

Printed in Australia by: Southwood Press Pty Ltd
80-92 Chapel Street, Marrickville NSW 2204

CONTENTS

> With the Human Condition resolved it becomes possible for the first
> time to reveal the real developments that humanity underwent
> throughout its 'adolescence' or search for meaning and identity —
> the psychological and intellectual adjustments we were making in
> response to the pressures of the Human Condition, in other words a
> description of the growth of what we now call our mind. To explain
> this, the psychological state of a teenager, a twenty-year-old, a thirty-
> year-old and a forty-year-old are described to illustrate both an
> individual's progress through life and humanity's progress through
> its adolescence. During the course of this description some major
> mysteries that until now have confounded humanity will be solved.
> Most importantly, the prime-mover or major influence in the
> development of humanity will be revealed and with it a reason for
> why (and therefore when) we began to walk upright, why (and when)
> we lost our body hair, how we became meat-eaters and how exactly
> we are related to the other primates. 'Sex,' I.Q., corruption, inno-
> cence, science, religion, prophets, men, women, animals, idealists,
> realists, have all played different, important and often unrecognised
> roles in humanity's development. What these have been and their
> interdependence are now explained.

PART TWO

The Unevasive Scientific Story
of the Ascent of Humanity

Part Two is broken into several sections. First the trinity of influences — namely the meaning of existence and the two biological tools by which it was possible to comply with it — are explained, followed by a description of what the tools achieved, their limitations and how these limitations were overcome. In the course of this description, many human behaviours are explained. (For ease of reference, boxes have been placed within the text which mentions the phenomenon being explained.)

The sequence of major events in the development of matter on earth are as follows:

A period in which we set out from paradise to find understanding of ourselves. It was during this period that the Human Condition, the state of conflict between human mind and conscience, emerged; a conflict we have courageously endured for two million years while we valiantly sought enough understanding to make resolution of it possible.

PART THREE

Conclusion

With reconciliation of our mind and our conscience possible, humanity is free at last. This section describes how our liberation has been achieved and what it means for us.

Summary

WHY HAVE HUMANS been individual, competitive, egocentric and aggressive? What is the explanation/defence for ourselves? What is the reason/justification for our apparently divisive condition?

The following is a eleven-paragraph summary of the reason. This brief summary was written for use as an advertisement to promote this book. It serves here as a preface to the book.

Assume just for a moment that our aim is to consider the welfare of the group or larger whole above self. In other words, rather than be selfish which is divisive our aim is to be selfless which is integrative (integration as defined in the *Encyclopedic World Dictionary* is 'the act of bringing together [parts] into a whole'). Further, assume that we have in us a genetically based instinctive self trained in altruistic behaviour but this training is only an orientation to selflessness not an understanding of it. Suppose we subsequently acquire an intellectual capacity to operate from a basis of understanding. What happens when we begin to attempt to manage life from a basis of understanding, given the already established instinctive orientation to selflessness? What happens when the mind meets the instincts?

The Birthday Party

To see what happens, imagine a children's birthday party where all the children sitting around the table are about seven years old except for one who is about eight years old. The seven-year-olds are still obedient to, and dependent upon, their instinctive training for management of their lives. The eight-year-old however has become sufficiently mentally developed to decide for the first time in his life to attempt to manage his life using his mind. Being an understanding device his mind requires understanding but there is none available to help him make decisions. Having to make a start somewhere on this process of thinking for himself he looks at the birthday cake and feeling hungry he decides 'well, why not take the cake'. But when he leans across the table and takes it the younger children, obedient to their instinctive training in selflessness and unaware of the misjudgments and misunderstandings which can be made as the mind searches for reasons for everything, criticise him for being selfish. (Interestingly, many mothers actually witness this grand mistake of pure selfishness often made by children first attempting to self-manage. Our word for such totally unknowing mistakes is naughty). So what happens is the eight-year-old gets a nasty shock and quickly puts the cake back on the table, determined not to have anything more to do with attempts to self-manage his life. The trouble is he cannot deny his intellect and so sooner or later has to find the courage to shoulder the responsibility of learning to master his mind. Again he takes the cake only this time he adopts the more subtle form of selfishness called reciprocity, where he offers the others some of the cake to keep them quiet. The instinctive selves of all concerned however are not deceived and continue to criticise him.

To continue his search for understanding the eight-year-old has no choice but to defy the criticism. First he tries to explain that he is not deserving of criticism — that he is not bad — so he says, 'Mum, the cake accidentally fell into my lap.' With this explanation he is evading the apparent but false implication that he is being selfish or bad. Lacking the real explanation for his behaviour his attempt at an explanation is inadequate, an apparently blatant misrepresentation. To be able to adequately explain himself he would need all the understanding that he is setting out to find, specifically an understanding of the difference between the genetic-based learning system that gave the innocents their orientation to selflessness and the mind-based learning system which requires an understanding of why selflessness is meaningful. He is going to have to discover and learn about the DNA molecule and the nervous system and many, many other mechanisms before he will have the means to free himself from criticism, a task that has taken the combined effort of all humanity some two million years to achieve.

When he cannot explain himself satisfactorily the eight-year-old becomes frustrated and tries to demonstrate that he is not 'bad'/'no good'/inferior/worthless. He hurls his chair away and challenges the innocents to throw theirs as far. When this sad and desperate effort to demonstrate his worth fails to impress he retaliates against the unfair criticism. He leans across the table and punches one of the innocents in the mouth in an attempt to shut up all the critics. Finally he tries to escape the unfair criticism: he puts his fingers in his ears to block it out and he runs away from the table to hide from it.

So the boy who set out in search of meaning/understanding became egotistical (forever trying to explain/prove/demonstrate/maintain his worth/'goodness'/self-esteem), competitive, aggressive, mentally 'blocked out' or evasive or alienated (paradoxically he was especially evasive of integrative meaning because of its unfair criticism of his divisive behaviour), escapist (superficial) and very unhappy. In short he became upset and because these upsets were mostly selfish rather than selfless traits his instinctive self criticised him all the more, compounding his upset and increasing his need for courage and determination to find understanding and now also to establish his worth.

The innocents and the eight-year-old's own innocent instinctive self saw his upset as proof of his 'badness'. From his first unknowing mistake developed what the innocents (including the innocent in himself) saw as deliberate mistakes. As far as they were concerned he had become 'evil' or 'sinful' although in reality it was their ignorance (of a mind's need to find understanding) that led to this intolerable and maddening situation. The eight-year- old was being forced to live with a sense of 'guilt' that at base was completely unjustified and unwarranted.

The origin of all the human upset on earth was our instinctive self's unjust criticism of our mind's necessary efforts to find understanding. From the time our intelligent mind emerged some two million years ago it has had to defy/resist/fight/battle/survive the ignorance of our older and more established instinctive self or soul, the expression of which is our conscience. If our conscience had had its own way we would never have experimented in self-management and so would never have found understanding. (Of course, while we had to be free from the oppression/restraint/criticism of our conscience to experiment in self-management we had to avoid diverging too far from its integrative ideals — too much freedom and we would become too upset or 'corrupted', too little and our necessary search for understanding would be oppressed.)

So while humans have appeared to be divisive or disintegrative, the full or greater truth is that we were not; at all times we have been committed to integration. To find understanding and by so doing become successful, secure managers of life, as is our great potential, we had to defy ignorance, we had to endure becoming upset, we had to lose ourselves to find ourselves!

To end the upset on earth it is this paradox of being 'good' when we have appeared to be 'bad' which has had to be explained. **It is explanation that has always been needed on earth. Only explanation could stop the upsetting criticism, could mediate the situation, could pacify our conscience.** Specifically, we have needed the explanation as to why integration or selflessness is meaningful, why the genetic learning system, which we have called natural selection, could only acquire an orientation to integration (and how it came to do even this when the limitation of genes is that they cannot learn unconditional selflessness since a selfless trait self-eliminates) and how the mind differs in that it is a self-adjusting system, able to understand and manage events towards integration.

We have needed answers to pacify our upset — to allow us to think without encountering criticism. Certainly we have been able to temporarily abandon thinking and its consequences. We could temporarily <u>escape</u> our upset self and its world by abandoning them and seeking the shelter of a religion (where the ideals are preserved but not explained) or embracing the discipline/oppression of socialism (where anything but the social or integrated ideal state is denied) or adopting the ideal future, the so-called 'New Age' (where we <u>act</u> integratively), but ultimately the true path to our freedom — to a <u>real</u> 'transformation' of our selves — lies back through our mental evasions or blocks, lies in a successful <u>confrontation</u> with our adversary, ignorance, lies in the ability to think without encountering criticism, lies in the ability to think our split selves back together, lies in the ability to understand. Ultimately humans had to master, not avoid, thinking but to do this we needed to get up the explanations with which to think. The trinity comprising integrative meaning, the genetic learning tool and the brain learning tool, had to be explained. We have to be able to understand why we have been aggressive — why there has been terrible suffering on earth — why the inequality between people — why racism — why feminism — why 'sex' — why wars — why psychosis — why the devastation of our planet.

It is these answers, this profound solution, that this book supplies. It gives the unevasive but full and therefore safe/compassionate liberating truth about ourselves. While that may seem a preposterous claim it is nevertheless true. This summary demonstrates that the many limited or partial truths that we have had to evade or avoid because they were unfairly critical of us — especially integrative meaning or teleology — can now safely be acknowledged <u>as part of</u> the full truth. The final paradox is that the liberating full truth could only be found by accepting instead of evading integrative meaning. While science had to evade integrative meaning and concentrate on finding understanding of the mechanisms of existence that would make final liberation possible, ultimately it was exceptional innocence, the lack of the need to be evasive, which was required to synthesise the unevasive full truth from science's evasively presented insights and produce the liberation. It is the story of the emperor with no clothes. In the end a little boy had to break the spell of our self-deception, had to dissolve the evasions. This book is not a contrived defence of humans, such as Robert Ardrey's selfishness-justifying 'territorial imperative' concept or 'Social Darwinism's' selfishness-justifying 'survival of the fittest' concept or Sociobiology's selfishness-justifying 'selfish gene' concept. It is the true defence for our divisiveness. It is the Full Truth that we have been in search of and working towards for two million years. Humanity is Free At Last from criticism.

This book is the condensed version of a much larger 500,000 word book that has taken thirteen years to write and is currently (1988) being edited. To introduce the concepts to be presented and because of their perceived importance this condensed version is being published ahead of the full version.

PART ONE
The Human Condition

FOR MOST OF US our main personal concerns are how to be happy, how to feel good about ourselves and how to establish our sense of worth or find our self-esteem. Our wider concern is how to stop human upset, suffering, conflict and destruction on and of earth.

For a long time we have been able to do little more than resolve these problems superficially. Now patch up repairs — of ourselves and our earth — are no longer sufficient. We are fast approaching a state of complete exhaustion, both of ourselves and our planet. We need a profound solution, the profound solution, if there is one, and there is.

To find the profound solution we have to go back to the basic questions most of us long since gave up struggling with. From there we have to carefully unravel human upset from its beginning. This will take us back through the mists of time to our earliest ancestors. It is a thought journey which will require a little patience as there are many paradoxes to grasp along the way.

As to the truth of the explanations to be given it is worth noting that when Charles Darwin introduced the idea of natural selection, noted scientist Thomas Huxley said 'how extremely stupid of me not to have thought of that'. The explanations make so much sense that after reading them you similarly will be left feeling how obvious they always were. In fact you will probably feel as though you and all humanity have been brought out of a trance.

Having not read the explanation yet you may be incredulous. In a recent book titled *The View from Nowhere*, philosopher Thomas Nagal argues that some issues about our world (specifically the age-old problem of good and evil) are so complex that maybe our brains are just not made to get to the bottom of them. While this may be the prevailing view we might recall that before Darwin the picture he was able to explain so simply — namely the variety of life on earth — must also have looked all but inexplicable.

To assist the reader the explanation in Part 1 is presented in steps. These steps will lead to the answer to both our personal problem of wanting to know how to be happy and the wider problem of wanting to know how to end the upset on earth.

Step 1
The Need For a Profound Approach

The first step is to acknowledge that our personal problems are only a microcosm of humanity's problems and, since they have their origins in humanity's problems, will only be resolved when a profound solution to these greater problems is found. To illustrate this, if someone were to hit one of us, we could superficially right the wrong, obtain redress, by taking the assailant to court and having a fine imposed. That might resolve the particular situation but it doesn't solve the real underlying problem, which is human aggression. To find the profound solution to human upset and aggression we have to look for the fundamental or original cause. We will come back to our personal problems in the end but first we have to look at the larger situation.

Step 2
The Paradox of the Human Condition

Step 2 is to introduce us to the paradox of the human condition and the upset it has caused us.

There are many avenues into the heart of the problem we humans have been living with but the most direct is the one that begins with the fundamental question of what is the meaning of life.

We have always wanted to know the answer to this question. With it we would know what to do next, would know the right and wrong decisions to make in terms of it. For example, if the meaning of life were to go down to the corner shop we would know which decisions led to getting there and which led us away from the shop — we could tell 'right' decisions from 'wrong' ones. We would have a reason for deciding on one course of action in favour of another.

It is here, almost immediately we begin thinking, that we run into a problem — in fact <u>the</u> problem. There was a complication associated with the question of meaning and it was such a difficult complication that its effect produced our human condition of upset.

Our problem has always been not, as we like to think, with finding the meaning of life, but with accepting it. We intuitively know the meaning of life, but it has set us at such odds with ourselves that we have refused to recognise it. This is the immensely difficult complication.

It all comes down to what we term 'love', which is our everyday word for the act of unconditional selflessness. This selflessness, this concern for others, for the larger whole above self, is necessary to achieve and maintain the integration or combination or coming together of parts into a larger whole, to achieve order.

Most people accept without thinking that order is more desirable than disorder; we seek it in all aspects of our lives every day. We gain satisfaction, contentment and pleasure out of bringing order to some small quarter of our existence. Only physicists have agonised over whether order is a natural way of things; whether there is order in the universe. Now even they have scientific evidence for this truth, it is embodied in the law they know of as The Second Path of The Second Law of Thermodynamics, or Negative Entropy. In non-scientific terms it could be said that the law states that we constantly combine, or integrate, smaller things to make larger things. Scientifically, this explains how atoms integrated to form molecules, which integrated to form virus-like organisms, which integrated to form single-celled organisms, which integrated to form multicellular organisms, which integrated, or grew, to form specie societies (this is the stage humans are negotiating on earth at the moment — we are developing the specie society of humanity), and which eventually will form societies of all species (which, metaphysically speaking, is the time the 'wolf will lie down with the lamb') and beyond that will achieve the stable arrangement, or order, of all things (where there will be 'peace on earth and in heaven').

This is mentioned now only so that when it is said that the meaning of life is 'to love' the reader will know that later (in Part 2) it will be explained what the profound first principle physical and biological meaning of 'love' is. At this stage the intention is only to introduce us to the complication that gave rise to our human condition of upset and for this it is only necessary to nominate love (or unconditional selflessness or integration) as the meaning of life.

The great complication is that while humans do not always act lovingly (we are often selfish for instance, which is divisive or disintegrative), this doesn't mean that we are in conflict with the meaning of life, even though it certainly seems that way. The paradox of the human condition has been that, while the meaning of life is to love, when we have not been able to love (to be integrative) we have not been meaningless! How this could be is the question answered in this book.

We have had many highly refined (and therefore likely to be highly perceptive) views of ourselves — theological, philosophical and biological — which this book reconciles, so the following mention of 'God' should not be seen to indicate that this book is a disguised witness to some particular religious faith. These various views have each in their own way described the paradox we have been discussing. For instance, taking an illustration from the theological viewpoint, most religions have described this paradox metaphysically by saying that while God is love, God still 'loves' us when we personally are unable to love. The Christian religion, for instance, holds that God is merciful, that God loves us in spite of our so-called 'sins'.

Now to examine the paradox a little more closely. We humans made 'mistakes', we 'failed' to be integrative, but although it may have looked like it, these mistakes did not mean we were bad. Fundamentally, despite the evidence, we are not guilty beings, we are not evil. Adults view the mistakes made by children as being a necessary part of growing up. You could say the same of humanity's mistakes. The behaviour that resulted in us making 'mistakes', that saw us being divisive instead of integrative, was very necessary. We were carrying out experiments in self-management, learning, if you like, how to use the complicated tool which was our brain. It was vital for the ultimate development of order on earth that humans mastered self-management because only knowing management of the development of order of matter could possibly achieve the ultimate order, 'peace on earth and in heaven'. However, that process of learning to master self-management involved making mistakes.

This paradox — of appearing but not actually being bad (although still having to restrain our apparent badness as much as we could) — was difficult to see. Because we constantly appeared bad we often mistakenly assumed we were, which left us feeling depressed and unhappy. After a while we would come to grips with the paradox and recall the fact of our fundamental goodness, restoring our faith in ourselves, restoring our self-esteem, only later to fall back into the hole of doubt and uncertainty. Without a clear understanding of this paradox it was all too easy to fail to understand — to 'lose our spirit', as we described it. Life was a constant struggle to understand. In fact the knowledge that we were not fundamentally bad was so elusive we normally found it prudent to avoid thinking. As someone said 'thinking is dangerous because it leads you into downwards spirals of doubt'[1].

Seeing our 'badness', being insecure about the fact of our goodness, we have sought, throughout the ages, a clear/non-abstract/non-elusive, secured in first principle understanding of this paradox. (Metaphysical assurances such as 'God loves you', though comforting, did not explain why God loved us — did not give us the ability to understand/defend ourselves.) Deprived of clear understanding we ended up upset with ourselves and with others. We have termed this state of upset the 'human condition'. Its source has been our inability to clearly understand and explain that we are not actually bad beings.

The following stages of humanity, from our early ape ancestors to ourselves today, will be more fully explained later. For reasons which will also become obvious later they are identified by terms unfamiliar to the reader but which are more accurate descriptions than those used by anthropologists. To introduce them: Infantman was our ape ancestor. Infantman existed from twelve million years ago to five million years ago before developing into Childman whom we know of in the anthropological record as the australopithecines. The australopithecines existed

---

[1] Rod Quantock of *The Book Program*, a radio program produced by the Australian Broadcasting Corporation (ABC), as reported in *Sayings Of The Week* in the *Sydney Morning Herald* newspaper, July 1986.

from five million years ago to two million years ago before maturing into the variety of humans anthropologists call Homo which is us, intelligent self-managing but insecure Adolescentman.

Childman, oblivious of the necessity to master self-management, had not become insecure and, as a consequence, upset. Happy, unworried instinct-controlled Childman lived in what we 'remember' as paradise. It was Adolescentman, Homo, who courageously shouldered the responsibility of searching for his identity, for finding out <u>why</u> he made mistakes, even though it meant a journey away from paradise to do it.

The prime mover in human 'evolution' was not meat eating or tool use or language development, as has frequently (and evasively) been propounded, but what was happening in our mind. Anthropologists have postulated that the varieties of early man represented various divergent or branching developments. Now it can be seen that there was no branching — that one variety led to the next. There was only one major development going on, the development of the mind but we, unable to look at our psychological development, attributed all significance to everything but it.

Finding our identity, finding the secured in first principle biological explanation for ourselves, makes it possible for us to go back and unravel all the upset that has occurred since humans first became insecure. We have reached the end of the state of insecurity which was humanity's adolescence and can now pass into adulthood, become Adultman where, as Martin Luther King said, we can be 'free at last'. Our freedom (from the human condition of upset) has arrived.

Step 3
The Story of the Birthday Party

Step 3 is to look more closely at how our upset occurred.

It began some two million years ago when Adolescentman found that his first tentative efforts at mentally or intelligently managing his life put him in conflict with his by-then-well-established instinctive controlling self. It was the start of a great battle which has raged until this day. On one side was our original instinctive genetic-based self that we have long known of as our soul and more recently have redescribed as our 'collective unconscious'[1], the expression of which we know of as our conscience. On the other, our developing mind or spirit. The problem, briefly, was that while our genetic self 'knew' that love or integration was the way to go, it did not understand <u>why</u> this was so, and our mind needed that understanding.

To see the battle it is necessary to understand the difference between genetic-based and mind-based learning systems. The whole problem of the human condition is resolved with this understanding.

Later (in Part 2) the limitations of these learning mechanisms will be described, along with the way in which these limitations were overcome in the process of developing the most possible order of matter on earth. That description is the story of the development of all the forms of life leading to the emergence of humanity.

So the story of the development of humanity involves a trinity of 'characters': the theme or purpose of our existence, which is developing order of matter, and the two great tools for doing it, namely genetic learning and nerve-based learning. (Incidentally, we have long been aware of the existence of this trinity of fundamental 'characters' or forces at work in our world. In Christian doctrine for example, they are described as God the

---

[1] Concept of Dr. C.G. Jung, see his book *Psychology of the Unconscious*, 1916.

Father, God the Son and God the Holy Ghost or Spirit. As we shall see, these correspond to integration, the instinctive expression or image of integration and the mind that searches for understanding of integration.)

For the present it is only necessary to look at one of the limitations of genetic learning. We know that natural selection, more properly referred to as genetic refinement, gave species the ability to adapt to new environmental conditions. In a long dry period, for instance, only long-necked varieties of giraffe, the ones which could reach the last, topmost, leaves, survived. In doing so, the giraffe species 'learnt' — it adapted itself to the new conditions. Genetic learning was a wonderful ability but it had this limitation: though change could be met, it could not be understood, and therefore it could not be anticipated.

However, a mind able to remember past experiences and compare them with present ones, associate the information, had insight and could learn to understand change. A mind was able to 'watch' what happened through time — was capable of becoming conscious of or understanding the relationship of events through time.

So, while Childman, Australopithecus, had been able (for reasons which will be explained later) to become instinctively or genetically adapted to being integrative, when the mind emerged it <u>had to</u> find an understanding of integration, it had to know why integration was important. This predicament has been recognised for centuries. For instance, it is perfectly described at the beginning of the Bible in Genesis where it says humans had already been made into or 'created in the image of God' (that is, they were integrative) but were yet to 'become like God knowing' (that is, understanding) (Gen 1:27 and 3:5). The problem was that, while our genetic self could tell us what was integrative or 'good' and what was divisive or 'bad' and in so doing guide us, it could not give us understanding of integration. We had to find that out for ourselves but much of our searching upset our genetic self. The effect of this was that our genetic self tried to stop the search. Our original instinctive self was <u>ignorant</u> of our mind's need for understanding and therefore unsympathetic towards the search.

Possibly the best way to illustrate how this battle and upset between the instinct-controlled self and the mind-controlled self emerged is to imagine a children's birthday party. All the children sitting around the party table are around seven years old except one who is about eight years old. The seven-year-olds are still obedient to their instinctively trained self while the eight-year-old is beginning to think for himself how to behave. Being an understanding device a mind requires understanding but there is none available to help the eight-year-old make decisions. Seeing the birthday cake and feeling hungry <u>he decides</u> 'well, why not take the cake'. But when he leans across the table and takes it the younger children, obedient to their instinctive training in integration, criticise him for being selfish. Many mothers have witnessed this grand mistake of pure selfishness often made by children first attempting to self-manage. Our word for such totally unknowing mistakes is naughty. Of course, on being abused by the innocents (the other children, who were innocent or unaware of the world associated with the need to search for understanding), the eight-year-old quickly learns the more subtle form of selfishness called reciprocity, where he offers the others some of the cake to keep them quiet. This however is only a superficial solution; the consciences of all concerned, including his own, are not deceived and continue to criticise any mistakes made in the effort to understand existence and self-manage on the basis of that understanding.

To stop the criticism the eight-year-old needs to be able to explain to the innocents and his own conscience that, while he requires conscience guidance, he does not deserve its criticism. He needs to explain that he is not bad; that he is using his mind, which is a different kind of learning tool to the one that gave the innocents their orientation to integrativeness. His mind requires understanding so he has to search for the correct understandings by trying or experimenting with different understandings. Any misunderstandings or mistakes are not bad but a necessary part of the learning process. However, to come up with this explanation the eight-year-old needs to know all that is being explained in this book about the difference between mind-based and genetic-based learning systems.

When the first Homo or Adolescentman or thinking man made his initial tentative experiments in self-management some two million years ago or more there were no such understandings and mediation available. He was just setting out in search of the understandings which have only now been achieved two million years later. It really was a Catch 22 situation: ideally to conduct the search for understanding he needed the understandings he was setting out to find!

The situation faced by the original Homos — and by the-eight-year-old — was that neither of them had an explanation they could use to defend themselves against the criticism from their original instinctive self even though they 'knew' it was unfair. Having to live with this unfair criticism — on a daily basis and through the millenniums — is what upset us. This upset took four forms. To use the eight-year-old's predicament at the birthday table, what he did when the innocents criticised him was:

Firstly, he tried desperately but unsuccessfully to explain or defend his actions. He said, 'Mum the cake accidently fell into my lap'. In truth this was not a lie rather it was an inadequate attempt at explanation. Like the birthday party boy, our 'conscious thinking self' (which is the dictionary definition of ego) became increasingly embattled and preoccupied trying to find proof of its worthiness. We became egocentric or preoccupied with self — selfish — which only upset our selfless behaviour-demanding conscience all the more. This compounding effect meant our upset intensified very quickly. In two million years of upset we have come a long way from the state of innocence and we have almost totally forgotten what true happiness is like. Humans have been immensely heroic but as a result we are now immensely exhausted. (It might be an idea here to briefly elaborate upon this egocentric factor as it explains our competitive nature. We have always 'known', although we have never been able to adequately explain it, that the greater truth was that we were not deserving of criticism, that fundamentally we were not bad. For this reason we would not tolerate criticism. We wanted it understood by others and we wanted to be able to understand ourselves why we were not deserving of criticism. Our struggle has been to explain and exonerate ourselves. Our lives have been focused on trying to prove that we were not bad. We wanted to win against the false accusation that we were bad. For two million years humans have been in competition with the implication that they were bad. For two million years we have been trying to win this fight — trying to liberate ourselves from criticism.)

The second thing the eight-year-old did was become angry. He leaned across the table and punched one of the innocents in the mouth in a frustrated attempt to stop the unwarranted criticism. This situation is the origin of our human aggression. Since aggression is divisive or disintegrative rather than integrative, this punch further offended the eight-year-old's conscience and fuelled his upset, as it did for humanity at large.

Thirdly, the eight-year-old tried to escape the criticism. He did this in two ways. First he put his fingers in his ears to block it out. Putting his fingers in his ears developed into more permanent forms of block-out when he mentally learnt to ignore the criticism by forgetting it, repressing it and evading it. In mentally hiding from or avoiding his real situation and adopting a false position he became psychologically separated from his true situation and self. This is the origin of our human alienation/psychosis/neurosis. Humans are now two million years psychologically blocked out or departed from their original self or soul.

The second way the youngster tried to escape the criticism was by physically running away from it. This is the origin of our human need for self-distraction which materialism served. The more we shouldered the responsibility of attempting to master self-management the more we were criticised the more we needed some relief, however superficial, from that unfair criticism. For instance we needed to pamper ourselves and go away on holidays. Work itself was often used to distract our mind.

It can be seen that, in order to gain the understanding that mind-based self-management was dependent upon, we had to live with the unfair — but necessary (for guidance) — criticism from our conscience. Throughout the search for understanding, the only ways we had of defending ourselves from this criticism or finding some relief from it were: by trying as best we could at the time to explain, prove and demonstrate that we were not bad, which was our egocentricity; by attacking the criticism, which was our anger; and by escaping the criticism, which produced our alienated self and our superficial, artificial, escapist lifestyle.

The battle left us angry, egocentric, alienated and superficial or, in a word, exhausted. We exhausted our capacity to be integrative. We spent the soundness of our soul. Inevitably our world became as false as we were. This was the price of progress (towards finding understanding). So, to answer our original question: the source of, the fundamental reason for the upset we all now suffer and which had to be explained to achieve a profound solution to our personal upsets, was this battle to overcome the ignorant criticism from our conscience of our mind's necessary efforts to master self-management. It was the uneasy transition from an instinct to a mind-controlled state. To use our religious or metaphysical terms, the situation was 'the origin of [so-called] sin'.

It might be of interest to include here a description written by the author Eugene Marais, the first person to conduct extensive field studies of primates. In his book, *The Soul of the Ape*, which was written in the 1930s but not published until 1969 he says:

'The great frontier between the two types of mentality is the line which separates non-primate mammals from apes and monkeys. On one side of that line behaviour is dominated by hereditary memory, and on the other by individual causal memory. . . . The phyletic history of the primate soul can clearly be traced in the mental evolution of the human child. The highest primate, man, is born an instinctive animal. All its behaviour for a long period after birth is dominated by the instinctive mentality. . . . As the . . . individual memory slowly emerges, the instinctive soul becomes just as slowly submerged. . . . For a time it is almost as though there were a struggle between the two.'

By 'causal memory' Marais is referring to the mind's ability to understand cause and effect which is its ability to 'watch' or learn about or understand what happens through time. By 'phyletic history' Marais is referring to the Phylum, the genetic inheritance or instinctive self. (The inhibition of conscious thought or effective reasoning in non-primate mammals and its liberation in primates will be explained in Part 2.)

Marais' books, *The Soul of the Ape, My Friends the Baboons* and *The Soul of the White Ant,* show him to be one of the exceptionally unevasive thinkers of this century. Robert Ardrey, who dedicated his book, *African Genesis*, to Marais, described him as 'a worker in a science yet unborn'[1]. Science as we have known it has had to be evasive or mechanistic, as will shortly be explained. Before unevasive science could be born, as it now is, the full truth that defends humanity had to be found.

To return to the explanation of the human condition. It follows from what has been said that the more we (intelligent man) searched for understanding the more egocentric, angry, alienated and superficial we became. In the end we arrived at where we are today — utterly spent, completely egocentric, angry, alienated and superficial. The more we tried to find understanding that would explain we were not bad, the badder we appeared to become! The truth is we have been enormously brave and heroic but this truth has been extremely hard to see. Unjustified as they were, guilt, depression, depleted self-esteem and psychosis have become rampant. To liberate us from this

---

[1] Description given on the back cover of the recent Penguin edition of *The Soul of the White Ant*, first published in 1937.

predicament we have needed to find the full truth about ourselves. Only with the full truth could our upset begin to be healed or rehabilitated. As the practice of psychiatry recognises, our freedom from our upset lies back through our mental blocks or evasions or repressions. To free our mind of its neuroses we have to 'climb onto the psychiatrist's couch' and try to confront, understand and so alleviate the repressed upsets/hurts in our mind. However our ability to do this has been extremely limited because humanity lacked the fundamental understanding needed, which was the defence for our exhausted, upset 'divisive' selves. Psychiatrists have had to be cautious about dismantling our mental blocks because they have not had the understanding with which to explain away the upsets that our blocks were hiding. Without the fundamental explanation for our upset human state psychiatrists were not in a position to begin to unravel the upset. Our neuroses were a bit like a locked room we could hardly afford to peep into. While we have long known the principles of psycho-therapy, as a practice it has remained undeveloped, even primitive. To quote a United States State Attorney-General, 'the art of psychiatry is just one step removed from black magic'[1]. Rehabilitation could not properly begin until the arrival of the full truth because we could not confront many of the apparent truths about ourselves until we could also defend ourselves against the criticisms implicit in them, since our defencelessness was what made us turn away from the truth in the first place. For the same reason, we could not recognise our embattled condition until we could explain it.

For instance, we could not confront our inability to love until we could explain why we had become incapable of love or unconditional selflessness. In the age-old 'nature/nurture' debate, for example, the argument has really been because the nurture thesis criticised parents when deep down we knew that we were

---

[1] The Attorney-General of the State of Massachusetts in the United States during the defence of a psychiatrist accused of negligence. Reported in *The Australian* newspaper, July 19, 1983.

not deserving of criticism and should not be made to feel guilty for our inability to love our children as much as we might have liked. Accepting the nurture thesis would leave parents unfairly and unbearably criticised, to varying degrees, so we have argued against the truth of it. (Of course while we were not fundamentally bad we had to restrain ourselves from, and in the extreme cases, punish excessive ill-treatment of children.) In fact the truth of the importance of nurture in our upbringing is only part of the greater truth, what could be called a partial truth. Like many other partial truths it was hurtful and dangerous because it left us unfairly criticised where the full truth would not. Discovery of the full truth, the reason <u>why</u> humans became upset and unable to love or nurture as much as they would have liked, would put an end to the debate by removing the need to evade the hurtful partial truth. Almost without exception those areas of inquiry that have remained controversial were those ones where exposure threatened — were areas where we were being confronted by a hurtful partial truth.

We have needed the full truth about ourselves before we could stop evading all the otherwise critical partial truths — before we could begin to be honest about or repent or own up to them, which is necessary if we are to clear up our neuroses/upset. Evasions were very necessary but they represented a denial — a repression of many truths. Evasions also upset the innocent among us who were trying to hold onto these truths and found themselves being forced into adopting the evasions. While evasion was necessary paradoxically it was also necessary to be as unevasive as possible so that we did not become too alienated from the truth. The finding of the full truth ends our need to be evasive and thus stops the spread of anger, egocentricity, alienation and superficiality within and among ourselves. We can now truthfully explain to innocents why we 'took the cake' which will end their upset with us and our upset with them for being upset with us. It is the freedom now to be honest that saves the world.

With sufficient explanation of why we did what we did, self-confrontation or so-called repentance or confession would not be difficult because necessarily there would be nothing we would be ashamed of — nothing we could not explain — could not actually or truthfully justify. What we have been seeking throughout the history of humanity was the ultimate excuse — was the excuse that was <u>the</u> reason and not the lie — and we now have it.

Our biggest evasion has been the way we have avoided acknowledging integrative meaning. For this, too, we first needed the full truth which explained our divisiveness, our inconsistency with integrativeness. We were God-fearing instead of God-confronting (remember integration depends on selflessness or love and God is love). In fact, while 'love' has been one of our most used words, science has not recognised it and has had no interpretation for it. Love means selflessness which is the basis of integration, the concept which we had to evade. The old Christian word for love was caritas which means charity or giving or selflessness (see the Bible, 1 Corinthians Chapter 13 and 10:24). As long as we evaded defining that what we meant by love was the tendency or desire to be integrative or selfless, 'love' remained a reasonably unthreatening word to use. We could not however be expected to be comfortable with the word 'integration' itself. In fact while we are familiar with the word 'disintegrative' we are relatively unfamiliar with its opposite or antonym 'integrative'. Integration and related words such as 'integrity' confronted us too directly with the truth of integrative meaning and, in the case of integrity, with our divided unsound self — with our lack of integrity — which we could not defend. We could identify with divisiveness but not with integrativeness. We did not like the word integration. We have evaded integrative meaning.

Science, instead of acknowledging a law of physics which says that systems grow by integration, acknowledged only an alternative law which says systems break down towards heat energy. We evasively identified with or preferred theories which suggested divisive, competitive behaviour was the norm, rather than theories of integration and co-operation. The position was such that a lie that said we were not bad was less of a lie than a partial truth that said we were! Such was the paradox of the human condition. Science had to evade the concept that there was purpose in existence, that there was a tendency towards order, towards developing larger wholes. Instead of recognising purposeful <u>development</u> it recognised only aimless change it termed 'evolution'. Science was mechanistic as opposed to holistic ('holistic' according to the dictionary means 'tendency in nature to form wholes . . .'). Science got on with the task of finding an understanding of the mechanism of change that might one day (as it now has) make possible a confrontation with the truth of integrative meaning — in the meantime responsibly evading such a confrontation.

In doing so we built up a veritable mountain of evasions which can now be dismantled. We became extremely false, choked up with evasion or denial, which was our psychosis.

Until the arrival of the full truth, the best we could hope for, both as a species and as individuals, was to delay the process of becoming exhausted. For instance, we could stop participating in the corrupting search for understanding for a while by isolating ourselves from the main battle. Or we could stop thinking (self-managing) and simply listen to and obey our conscience. This was the refuge religions provided because it was in them we enshrined the absolute truth of integrativeness that our conscience knew. (In fact the word 'religion' itself is an embodiment of the absolute truth of integrativeness being derived, as it is, from re-ligare which means 'to bind together', or integrate.)

Over the last three thousand years a few exceptionally innocent men have appeared who, through the rare circumstances of encountering only pure love in infancy and of being isolated from upset in childhood, retained access to their consciences. As we became more battle-worn and separated from our own consciences, as our need for refuge and temporary healing has increased, the more precious the unrepressed consciences, the words of these unevasive thinkers, these prophets or holy men, became. (Incidentally, the word 'holy' has the same origins as our Saxon word 'whole' so, like the word 'religion', 'holy' was concerned with our desire for wholeness, specifically for integrity or soundness of self, a self unseparated from its conscience, non-alienated. While integration of self and integration of matter are not the same thing, being able to be unevasive and recognise the development of the integration of matter depended on being integrated in self or non-alienated.)

Religions were one of the two ways we had to return to the world of our conscience and its integrative ideals. The other way was to simply deny the corrupted world of our mind and insist on adherence to the integrative ideals. We could adopt the politics of socialism. The words 'socialism' and 'communism' mean stress on being social or communal, on being integrative.

Whenever we became over-exhausted we could abandon our mind with its upset corrupt behaviour and be born again into the world of integrativeness — either by taking up a religion where integrative knowhow and ideals were preserved or by adopting the restraint of socialism where anything but the integrative ideal was denied. As Karl Marx said when describing the philosophy of socialism, 'the point is not to understand the world but to change it'. By 'change it' he meant make it social or integrated.

Life under the human condition was always a question of trying to balance the need for our mind to be free to search for understanding with the need for obedience to our conscience and its integrative or social ideals. Excessive freedom and we would become too upset or currupt; excessive obedience and our need to be free to search for understanding would become overly oppressed. The difficulty was that at any one time we never knew where the perfect balance lay. The only way we could find the approximate point of balance was to pursue freedom until it became obviously excessively corrupt then swing back in the other direction and pursue obedience to the ideals until that attitude became obviously overly oppressive, and so on, back and forth. In the extreme the conflict and argument over where balance lay could — and did — result in war.

While we could temporarily contain and even heal our upset or exhaustion in these ways we could not end it because, being a mind-based understanding device, the need for self-management would eventually reassert itself, at which time the march towards exhaustion would continue. Without the arrival of the full truth complete exhaustion was inevitable and in fact complete exhaustion has been fast approaching on earth. Our hope and faith that the full truth would arrive before we destroyed ourselves and our earth was being sorely tested. Only the arrival of the full truth could stop this progression and it has come only just in time. Our upset can now be pacified. The finding of the full truth will bring an end to argument, politics and wars. It will bring the human race together and it will bring our own divided selves together; it will make us happy. It is the realisation of all our hopes and dreams.

The slide towards complete exhaustion has been halted and we can now begin the journey back to paradise. The way forward now is back because going back is at last possible. We are entering the age of rehabilitation.

The fact that we are now safe (from self-destruction) will bring wonderful relief to us. The struggle and desperation has been so great that this sense of relief is possibly all we will be able to cope with for some time. When we recover enough to set out on the homeward journey we will discover it to be very different from the outward one, fraught as it was with constant battles.

Step 4
Rehabilitation of Humanity

Step 4 is to introduce us to life on the homeward journey.

It will be, relative to the two million year outward journey, a very quick trip lasting possibly only ten generations and the progress each generation makes will be immense. Even today's generation will take great strides simply by experiencing the relief of turning around and facing in the right direction for the journey following generations will make.

Nevertheless, this journey will have its own particular problems. The main one will be that while we are exonerated we are also exposed to the extent of our exhaustions. While at the broad level, humanity is incredibly battle-fatigued, individually we are each at different stages of exhaustion. Some of us are just going into battle; some are in the midst of it, some are retiring exhausted and some are recuperating making ready to go back. While all these different stages of exhaustion can now be relieved, this process will take time during which we will remain exhausted, to varying degrees. The problem is living with our exhaustion while we are still exhausted.

During the generations it will take to heal our ego, anger, alienation and superficiality, the extent of these upsets in us will be exposed. When we can see how we could be we will not like being exhausted and will not like others knowing the extent of our exhaustion. This is in spite of the fact that our fatigue is now totally justified — completely defensible. While we are truthfully defended now it will take time to let go our old false defences and adopt the new truthful defence. In short we will remain insecure during the time it takes us to become secure.

For some time to come we will continue to misunderstand and consider those who are innocent or unexhausted as being 'good' and those who are no longer innocent and now exhausted as being 'bad'. Similarly we will find it hard to adjust to the fact that our own personal exhaustions are not bad. It will take time to learn to love ourselves and others. It will take time to overcome our insecurity and take the freedom that is now available to us. It will take time to realise the door of our jail is now open. Further, we will discover that it is one thing to understand and still another to know and trust with our whole being. We will discover that knowledge follows many months behind understanding.

On the outward journey we were able to evade or block out the extent of our battle-fatigue or corruption; on the return journey we won't need to and won't be able to but we will still be inclined to want to hide. We will procrastinate and cling to our old evasions but they will no longer work. Adopting the new truthful defence for ourselves will be a radical re-adjustment we will find difficult to make even though our old defences are now transparent. To some extent we will be stranded between the old and the new world. The novelist Gertrude Stein coined the phrase 'the lost generation' and the reason the phrase caught on was that we knew it had significance. As can now be seen there will be a stranded lost generation or two, however even this is nothing to despair about as there will be many compensations for these generations — such as the euphoria of the homeward bound journey (as will shortly be mentioned).

Up until now we have survived by hiding. Our whole existence has been built upon developing artificial defences, evasions, props and disguises for ourselves. These are now destroyed and replaced with the true defence for ourselves but the problem is we need time to adopt this new defence. It is a shock. This is the real 'future shock'[1] that we have suspected lay ahead of us. Our biblical term for it was 'judgement day'. Judgement day is exposure or revelation day. We have finished with the age of guilt and we now pass through a brief age of exposure before entering the age of freedom. The best way to get through the age of exposure is not to procrastinate but to get on with it and get it over with.

To illustrate the problem: when it is explained that nurture not nature is all important in bringing up our children many parents will still feel they are being unfairly criticised and will go back to our old false evasive defences and say: 'No, we are not guilty because it was nature (our genetic makeup) rather than our nurturing that was all important'. Similarly we will go on trying to deny integrative meaning to falsely justify our divisive

---

[1] Reference to the book of this name (published 1971) by Alvin Toffler.

competitive ways and, continuing to believe our exhaustion is bad, we will go on refusing to accept it, denying it, pretending it does not exist. We will try to maintain our old ways of coping. We will continue to be evasive or false even though we no longer have to be false and are aware that by continuing to be false we are perpetuating falseness and the upset on earth. We will procrastinate. However, we can and will gradually learn to adopt the new truthful defence for ourselves and say: 'My children are upset because I was not able to nurture them — in the same way I was upset because my parents could not nurture me — but the good reason for this is that I and my ancestors have been heavily involved with humanity's necessary battle to find understanding', or 'I am alienated and superficial because I have been battling to find understanding for humanity'. As soon as we say this, as well as ending the criticism from others who were upset by our evasion, we have also taken a giant step towards ending our 'inability to love our children' or towards ending our 'alienation and superficiality' — towards ending our upset with the criticism we were receiving from those who had yet to go to battle. The full truth is the oil we needed and, now that we have it, must pour on the troubled waters of our time to still them.

We have to learn to explain that there are no such things as good and bad people only people who are living with different degrees of exhaustion from fighting <u>for</u> humanity. We have to remind any who persist with the notion of 'badness' that at the end of a football match the battered and bruised heroes who are carried off the field high on the shoulders of their team mates are the players who did all the work in the thickest part of the match. Our exhaustion, like that of the football players', is admirable, not despicable. The boy who had the courage to search for understanding, to take the cake and battle the unfair criticism, was the hero not the villain at the birthday party. For two million years our soul/innocence has misunderstood this but now we can clear up its misunderstanding and our upset with it.

Abandoning our old false defences for the true ones will be difficult at first but once underway not as difficult as initially thought. Proof of this is that humanity has been 'coming out of the closet' — owning up already to such behaviours as 'sex', incest and the extent of child neglect — even without the safeguard of the true defence for these upsets. If we could confront our upsets when it was so threatening to do so how much easier will it be now that it is safe. However — a note of caution. While sooner or later it will be necessary for each of us to bravely make the first awkward effort to adopt the new defence for ourselves ultimately we will require the help and guidance of specially trained people. Until humanity has mastered the skills that will be necessary we should limit our efforts to self-expose and psychologically analyse others to what comes naturally to us as our minds absorb understanding. There are many, many completely new understandings (the dismantled 'mountain of evasion') to become familiar with before we can claim any degree of skill in understanding and explaining away our evasions, confusions and upsets. The door to the new world has only just been unlocked which means there is a lot to discover, learn and adjust to. We are now at the foot of a very steep learning curve.

These then are some of the problems of life on the homeward journey. However, there are many exciting positive aspects to compensate for these negatives. The greatest is the fellowship that is now possible on earth. While we will still be upset or battle-weary during our rehabilitation we can at last see the lights of home and the joy and excitement of this sight will bring us together in spite of ourselves. For instance, now that we can see how extremely upset we really are and how safety is at last within our reach we will want to call a moratorium on our now baseless personal and international angers. All we have to do now is buy time to absorb understanding. We will want to call a moratorium on humanity's capacity for nuclear, chemical, genetic and technical destruction. To be acting like God while we could not even confront God was extremely dangerous. What was even more dangerous was that our upset could become greater than our capacity to care with the consequent indifference to the future of humanity. This can change now. Our hopes are now realities. Enthusiasm is going to come thundering through and cynicism is going to die. Everywhere people are going to come climbing out of their trenches and start hugging each other. We have won our two-million-year battle to get the truth up.

With the answers we can free ourselves from upset. Freedom is not here yet but it is in sight — it is at last possible. We can know that we can make it where before the situation looked all but hopeless. After our long and terrible journey through the darkness and wilderness we are necessarily all in tatters and utterly exhausted but seeing the lights of home at last we will find the energy to help each other through the final few miles of entanglements. Nothing can stop us now. The gate is open to lush pastures, streams and fields of flowers and there is no point staying on in the barren desert trying to make the most of the few paltry sticks we had to play with out there. There is a world that is so huge in its magnificence — in its excitement — in its happiness — in its wonder, that all our sophistications developed to the utmost to keep us going through the agony of our adolescence (such as 'one dozen oysters natural' and 'the enjoyment of a cup of coffee in a fine china cup after a lovely evening meal') rate as <u>nothing</u> beside it. (Shortly in Step 5 it will be explained that we have had to repress our memory of paradise because its beauty only reflected critically on our battle-weary state that we were unable to defend. We could not afford to show feeling and the more embattled we were the more this was so — the stronger we had to be. To quote John Lennon from his song *Nobody Told Me*: 'everyone is crying but no one makes a sound'. For the first time in two million years it is safe to show emotion and to recall the beauty and happiness of our lost pure world and it is important to do so to inspire our homeward journey.) Everything is now possible. There is a lot to do but what is so wonderful is it is all now possible. Soon from one end of the horizon to the other will appear an army in its millions to do battle with human suffering and its weapon will be explanation which is understanding.

Thinking can now begin in earnest. We can start back down the dark corridors in our brain unlocking all the doors, dismantling all the blocks in our minds. The unravelling process can begin. We can think ourselves back together. We have to go back to all the things that went wrong in our lives and replace our misunderstandings, upsets and evasions with understanding, compassion and honesty. With understanding upset goes.

We can come out of the caves we have had to live in to hide from the glare of the many partial truths that used to hurt us and sit in the sunshine of understanding and start thawing out. Instead of hiding and repressing we can now start seeing and revealing. We no longer have to suffer further upset <u>and</u> we can begin to unravel and heal the upset that has occurred. In the past the only way we could cope was by repression, escape and attack, which led to further exhaustion. Now at last we can cope by understanding.

We have to appreciate that at the beginning of the homeward journey all that will be different in terms of what has happened to us is that we will be facing towards home instead of away from home. We have, finally, changed direction. Rehabilitation of our upset is now possible whereas before it was not. Yesterday we were exhausted and becoming more so. Today we are still exhausted but becoming less so. While initially all we have done is change direction it will make a big difference to our lives. Instead of deteriorating our situation will begin to improve. Our mind will take time to absorb the answers, unravel the confusions and upsets and heal itself. This cannot happen suddenly but great changes will take place, even in one lifetime. Our personalities are the expression of our various confusions and upsets; they won't change overnight. However one day after about a year of living with the explanations each of us will look back and suddenly see that we have changed a great deal. A lot of upsets will have died down and even vanished. We will see how we have been rehabilitating or healing. This process is not an abandonment of our mind's worries, our upset self, but an acknowledgement and understanding of the pain that we incurred and the reasons for it, which dissolves the pain/upset.

In contrast, religious conversions, where we abandoned our mental upset self in favour of living through Christ (or one of the other prophets the world has known) — and in so doing be 'born again' as a force for integrativeness — brought instant relief from our upset but did not resolve it. Religions offered 'salvation from sin', they offered a way out or escape from our upset but they were not able to liberate us from upset by resolving it. Ultimately humans had to learn to master thinking, confront our exhaustions, and in so doing resolve the upset. Our true freedom lay back through our blocks not further away from them. As an illustration of the instant and even miraculous relief religions offered, Pat Robertson, one of the leading television preachers in the U.S.A. described his conversion experience thus: 'At my desk in my office, I leaned back in my chair and burst out laughing . . . I had passed from death into life.'[1] This relief — this 'salvation from sin' — achieved through such rejection or abandonment of our upset self to a faith is no longer necessary, instead we can think ourselves back together, become genuinely whole again. However, religions had an important role to play in the past and undoubtedly will still play a large part in supporting many of us for some time to come. We are immensely insecure and many will need the help of our religions until we become stronger in our ability to defend ourselves truthfully — to think without encountering criticism/pain. The thing is, thinking is now possible whereas before it was all too often hurtful and dangerous.

Everything will become positive now where before it was negative and draining. In truth we were not living we were dying. While we remained physically alive our soul died because we blocked it out because it criticised us and our mind or spirit died because thinking became too dangerous and its effects too destructive. With understanding now available this can all

---

[1] *Time* magazine, February 17, 1986.

change. The most dramatic result will be with our children. Children come into the world with relatively pure souls, instinctively expecting an innocent world, a world without upset. In the past this meant that when they encountered our upset adult behaviour they found it extremely distressing. To their innocent consciences adult behaviour was all so terribly wrong. As Antoine de Saint-Exupery said in his famous book, *The Little Prince* (1945), 'Grown-ups are certainly very, very odd'. This very, very odd world that children found themselves in would not have mattered so much to them if parents had been able to tell them why it was so odd. It was not what happened in our lives that was the problem so much as our inability to understand why it was happening and so cope with it honestly. It wasn't the taking of the cake but our inability to explain why we took it that was the problem. Without explanation to defend their various adult upsets parents could only deny them, give false explanations for them or simply be silent about them. Kept in the dark about what was going on children were left with no alternative but to block out the pain the adults' upsets and strange world caused them. In this way, since human upset first appeared on earth, the members of each new generation have had to learn to repress their true selves and adopt the prevailing levels of evasion, denial and silence on earth. Now, suddenly, this pattern is broken. Now with explanation at last available children will no longer have to die inside themselves in a sea of silence, superficiality and what is to them lies. Children will be able to be told why we are the way we are.

This means they will stay alive inside themselves and we will soon see adults appear having all the happiness of young children. The strength they will derive from that happiness will absolutely obliterate any problems that remain. This is where that army it was said would appear on the horizon will come from. Compared to our task of struggling with an intolerable burden of unfair criticism the task of these later generations will be easy. Those who lived during humanity's two million years of defenceless adolescence where the whole world in effect disowned them for their unavoidable divisiveness are the truly great heroes. We and those before us have been the ones who had to overthrow ignorance. The picks that demolished the granite wall that stood between humanity and heaven will be the most cherished objects in all of heaven. This earth has never seen and will never see again anything so great as us. We have been incredibly heroic and when we come home all nature will line the streets to welcome us. Weary and in tatters the victorious army of humanity has finally broken into the kingdom of heaven. We have still to realise it but we are now standing <u>inside</u> the gates of paradise.

Step 5
The Unevasive or True Story of the Life of a Human During Humanity's Adolescence

The whole story has now been told. What remains is to fill it out. An effective way to begin this might be to take up the story of the eight-year-old boy. This is a legitimate comparison. Since the prime mover or main influence in the maturation of both an individual human and humanity as a species was what was happening in the mind, both underwent the same maturation stages. The reason it is the male's position that is described is that the priority concern throughout humanity's adolescence was this spiritual or mental fight/battle to defy/overcome the external threat to humanity of ignorance which, as group pro-tectors was a responsibility that fell to males. Since this battle to defy ignorance was humanity's priority concern it meant women had to help men. Women were 'man's helper' (Gen 2:18). Everyone and every other need was subordinate to the priority concern of winning this battle. We were a patriarchal or male role prominent or dominated society. It should be stressed immediately this does not mean the recent feminist movement was not legitimate and necessary. It was, as will shortly be explained.

We left the boy at the birthday party making his first experiments in mind-based management. He has already discovered that self-managing when there are no understandings or answers is not at all easy. Sadly from there things only become worse. What ushers in his adolescence is the sobering realisation that the full implication of trying to manage life without understanding means a life of complete chaos and upset. He has to resign himself to this fate; he has to find the courage to accept what he cannot escape. From here on, more and more courage will be asked of him. From a happy child he becomes a sobered adolescent. Most unhappily of all he cannot discuss his problem with others. Unable to defend the battle we were participating in none of us could afford to admit to it, so we could not even talk about it. We could offer each other only sympathy, superficial comforts and possibly a distracting and mood-lightening story. At about the age of sixteen the adolescent is at the height of his silent battle to come to terms with his fate. It will take him another five years of internal wrestling to finally climb on top of his initial depression and realise that, well, while he will undoubtedly end up defeated, at least he has the adventure of the battle to look forward to.

By the time he turns twenty-one he has long since learnt not to think about the prospect of his inevitable defeat. He has developed a totally 'positive' attitude, which really means he has learnt to block out the negative truths. He has been able to arm himself sufficiently well with evasion, with a positive attitude, to commit himself to the battle. In fact he has done such a good job of blocking out any negatives he is raring to go. He reasons that while he might eventually go under he is determined to make a good fight of it. He is cavalier and swashbuckling. He has plenty of strength and resilience — plenty of rock and roll.

What has really happened is that, through extreme resignation, his life has been reduced to looking forward to the excitement of the adventure entailed in the journey through life under the oppression of the human condition. 'The adventure' of this journey is seeing exactly how the battle will unfold for him. He might know that he will be defeated but he does not know how that defeat will take place. What is in truth a very small positive has, through sufficient blocking out, been made into an all-consuming wonderful positive. Such is his courage — but it has taken twelve years of mental effort on his part to make these necessary but horrible preparations. For instance, he has had to block out possibly 70 per cent of his awareness of the beautiful world he inhabited as a child. To have to 'see' the beautiful world that he can no longer be a part of would make his life unbearable. (Of course we have always to remember that the greater truth is that while he is actually going to his soul's and mind's death he is doing so in order that, one day, the beautiful world might be permanently restored to everyone. Humanity had to lose itself to find itself! In the next 20 years of battle the remaining 30 per cent of these memories will also be obliterated and he will be left walking in a terrible darkness. This is his real prospect. The courage of humans who had to face it has been so immense it is something that is, and possibly will be for all time, out of reach of appreciation.)

In the story of the maturation of humanity, the equivalent of the sobered teenager who lived through the resignation stage of adjustment to the human condition was Soberedman, the first of the Homos anthropologists know as Homo habilis; he lived from two million years ago to one and a half million years ago.

But we were up to the stage where the boy has become a young man of twenty-one about to set out on his 'life's adventure'. With a big kiss from Mum and a slap on the back from Dad he leaves home 'to see what life holds for him'. From a sobered teenager he has become an adventurous twenty-one-year-old.

His initial task of self-managing his life has long since been transformed into a battle to establish his worth — to establish that what he is doing is not bad. Ultimately the only way this can be done is by finding the first principle biological explanation of the difference between a mind-based and a genetic-based

43

learning system — a task which will take the combined effort of all humanity some two million years to achieve. So all our twenty-one-year-old can hope to 'achieve' in his single lifetime is to keep the effort going and possibly contribute a few clues towards the finding of the full truth one day in humanity's future. While this is the larger view of his life the view from his position is, as mentioned, that he wants to establish his worth. He wants to express himself — to satisfy his ego — to keep his self-esteem intact as much as possible. To do this he has to find whatever ego reinforcement he can from whatever situations he encounters.

In our evasive jargon he has to 'achieve as much as he can'. As well he can try to repel any implication that he is not worthy by aggressively counter-attacking the quarter from which such an 'attack' on his credibility comes and/or by blocking the criticism from his mind. When these means for propping up his ego begin to fail, as they must eventually, he resorts to trying to escape from the battlefield. At this stage, when battle fatigue has set in, he starts to feel the need for material rewards and distractions. (Unable to recognise exhaustion as admirable in the past, material rewards became our way of having the honours due to us for our courage and effort which led to our exhaustion bestowed upon us. For example, when we were exhausted we could dress in glitter and have huge chandeliers in our house to give ourselves the fanfare we knew was due us but the world in its ignorance would not supply.) From being confrontationalist he becomes escapist and increasingly superficial. He abandons any hope of winning and now concerns himself only with finding relief and bestowing glory upon himself.

This has taken the story a little ahead of itself. The young man's twenties and early thirties are mostly spent refining the devices for coping. He settles into life under 'the human condition'. Our ancestral equivalent is Adventurousman, whom anthropologists know as Homo erectus. He lived from one and a half million years ago to half a million years ago. It was Homo erectus who adventured out from our ancestral home in Africa around one and a quarter million years ago. During his one-million-year reign Homo erectus mastered the many techniques for coping with the human condition which we now take for granted and euphemistically refer to as 'human nature'. It was Homo erectus who perfected what we know of as the hunter-gatherer lifestyle.

Contrary to accepted opinion, the hunting in the hunter-gatherer lifestyle was not for food. This evasive belief has so far protected us from the critical partial truth of the extreme aggression involved in hunting. In fact research shows that 80 per cent of the food of existing hunter-gatherers, such as the Bushmen of the Kalahari, is supplied by the women's gathering.[1] If providing food were not the reason, why then were the men hunting?

Hunting was men's earliest ego outlet. Men attacked animals because their innocence, albeit unwittingly, unfairly criticised them. By attacking, killing and dominating animals men were demonstrating their power, which was a perverse way of demonstrating their worth. If men could not rebut and thus win against the accusation that they were bad at least they could find some relief from the guilt engendered by demonstrating their superiority over their accusers. The exhibition of power was a substitute for explanation. This 'sport' of attacking animals, which were once our closest friends, was the first great expression of our upset. Anthropological evidence suggests that large-scale big game hunting began during the time of Homo habilis and became well established early in the reign of Homo erectus. With big game hunting came meat eating, which would have revolted our original instinctive self or soul since it involved eating our soul's friends. But our spirit wasn't to be put off and in time, as we developed our increasingly upset and driven (to find ego relief) lifestyle we became somewhat physically dependent on the high energy value of meat. Nevertheless, while using the meat for food could justify the effort of hunting, the hunting was

---

[1] Kalahari *Hunter-Gatherers* by Lee and De Vore, 1976, Page 115.

really all about the boy leaning across the birthday party table and punching the innocents in the mouth.

Next men turned on the unwitting but critical innocence of women and destroyed that. Men invented sex, as in 'fucking' or destroying, as distinct from the act of procreation. What was being 'fucked' or destroyed was women's innocence. To explain as briefly as possible how this happened: It was the arms-free facility of primates that made it possible for them to hold and carry a helpless infant and in so doing allowed the development of an extended infancy period. An extended infancy period meant that infants could be exceptionally trained in selflessness or love. While 'mother's love' or maternal selflessness was actually genetically selfish (as will be fully explained in Part 2, a limitation of the genetic learning tool was that a genetic trait had to reproduce to survive) it appeared as selfless behaviour to an observing brain, so by prolonging infancy an infant's brain could be trained or indoctrinated in love, a process which throughout this book is called 'love-indoctrination'. The 'trick' in this process of 'love-indoctrination' was that it allowed selfless or co-operative or integrative behaviour to be learnt in spite of 'gene selfishness'. The unconditional selflessness required in developing an integrated whole or group could not be learnt genetically except through this remarkable process of love-indoctrination. While in fact the mother is genetically selfishly looking after her own reproduction by looking after her infant, in appearance maternalism is unconditionally selfless behaviour on the part of the mother towards her infant, so all the infant experiences and is thus trained in is pure selflessness. Science being evasive of integrative meaning could not recognise this integrative process of 'love-indoctrination' and actively evaded it, claiming maternalism was nothing more than a mother protecting her helpless infant. We have had to live with the quite unbearable fact that during the two million years of humanity's adolescence we have been unable to love our infants as much as we were able to during humanity's infancy and childhood. The way we made it bearable was to evade the whole significance of nurturing in our history and in our personal lives.

The recently advanced theory of sociobiology sought to excuse divisive human behaviour on the grounds that it was due to the genetic need to be selfish. This was one of our latest scientific evasions. The unevasive truth is that one of the integrative limitations of genetic learning was that genes had to be selfish. The significance of love-indoctrination was that it allowed integration to be learnt in spite of this limitation.

It was through love-indoctrination that we acquired our original instinctive training or orientation in love mentioned earlier. It was the way our conscience acquired its orientation to integrativeness. Over many thousands of generations our maternal training in selflessness became genetically reinforced, it became an instinctive expectation in us. (Although it has to be remembered that while we became instinctively trained or indoctrinated with love, while we learnt how to love, significantly, we did not learn why we should love.) Initially the way the process of love-indoctrination was genetically encouraged was by 'naturally selecting' more maternally capable mothers; once initiated the genes could then reinforce selfless training. On their own genes could not develop selflessness but once there was love-indoctrination to train individuals in selflessness the genes could then follow the training process, reinforcing it. Later, when the mind went its own 'corrupted' way during humanity's adolescence, the genes would similarly follow along behind it, as it were 'reinforcing' what was happening. If sufficient generations of humans decide to want to hunt animals gradually there will occur genetic selection for and thus adaption to hunting. Genes would naturally follow and reinforce any development. The difficulty was in getting developments to occur, not in making them instinctive because that was automatic. In Part 2 this process of love-indoctrination will be looked at again within the full biological description of our development.

In the case of the instinctive reinforcement of love-indoctrination eventually self-selection was to play a part in leading the genes. We came to recognise and seek as mates individuals who would make our group more co-operative (or integrative). The right sort of individuals were those who had spent the longest period in infancy (and therefore received the longest training in love or selflessness) and those who were closest to their memory of this infancy training period. The physical features that signalled these traits were the neotenous, or childlike, features of large eyes, snub nose and domed forehead. Long ago we sought ('selected for', in Darwinian terminology) those features now regarded as cute and beautiful in a human.

These pictures of a juvenile and adult chimpanzee show the greater resemblance humans have to the baby and illustrate the principle of neoteny in human development. They appear in the book, *The Mismeasure of Man* (1981), by Harvard scientist, Stephen Jay Gould. Gould is one of the few scientists in the world who has been actively trying to resist and even expose the evasion of science. (In the case of *Mismeasure* he is referring to intelligence testing which was an evasive aspect of our academic system that will also be mentioned later in this book.) Gould's explanation for neoteny (where rates of development slow down and juvenile stages of ancestors become the adult features of descendents) is that 'We retain not only the anatomical stamp of childhood, but its mental flexibility as well' . . . that 'flexibility of behaviour are qualities of juveniles, only rarely of adults.' . . . that 'Humans are learning animals.' Gould correctly recognises neoteny as having played a significant part in our development. As Gould says, 'one picture is worth a thousand words'. However Gould incorrectly thinks the change occurred so we could learn. As has now been explained the cause of our neoteny was the process of love-indoctrination and the reason for love-indoctrination was to become trained in love or integrativeness. Neoteny did not occur so we could become 'mentally flexible' or 'learn'. Mentally clever man, Homo, appeared well after we were becoming neotenous. Neoteny occurred as a consequence of love-indoctrination and love-indoctrination occurred because

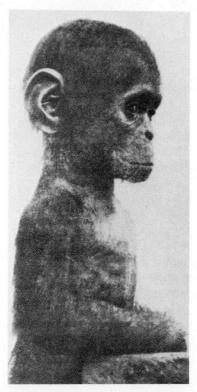

we needed to be integrative. It was integrativeness (co-operation) and neoteny that went hand in hand; our intellect developed later on. A fortunate but nevertheless accidental consequence of being trained in integrativeness by love-indoctrination was that our mind was liberated to think effectively. (As mentioned earlier, this liberation of conscious thought or effective reasoning in primates will be fully explained in Part 2.)

These two pictures are from an article titled *The Rare Pygmy Chimp* by Paul Raeburn which appeared in the June 1983 edition of *Science 83* magazine. The rounder eyes, smaller ears and less protruding jaw of the adult pygmy chimp (top) indicate that this species has developed more love-indoctrination than the common chimp (bottom). Although the researchers who wrote the article were unaware of the process of love-indoctrination, they said the pygmy chimps were frequently seen '. . . standing upright and walking on two feet [the more we had to hold a helpless infant with our arms during love-indoctrination the more we had to learn to walk upright]. They are known to share their food [they are trained in selflessness], and unlike common chimps, they frequently mate face to face [reflecting a greater awareness and thus capacity for affection]. . . . [Also, they] are much less aggressive than common chimps . . . Social groups are also more stable among the pygmies, who seem to get along with each other better [they are more integrated] . . .'. The authors also point to behaviour which suggests that females play a greater role in directing groups, in contrast with common chimp society, where the males dominate. (In the pygmy chimp matriarchy is replacing pre-love-indoctrination male dominance. The reasons for male dominance conventions will be explained in Part 2.) With one of the female pygmy chimps the authors found that 'you frequently have the feeling that she is trying to communicate about things in the past [an indication of the emergence of consciousness] . . .'

It is obvious that these rare pygmy chimps are going to be

absolutely invaluable for gaining an insight into humanity's infancy stage of development.

It can be seen that it was maternalism that made us human. Throughout humanity's infancy and childhood, a period that lasted from twelve to two million years ago, women's role of nurturing the infants was the most important one to the group. Men had to be in support of it. Women had to devote all their attention to the loving of their infants if integratively trained and thus co-operative adults were to occur. Until it became instinctive, love-indoctrinated co-operation was extremely hard to develop and maintain which, incidentally, is why there are many primate species still stranded in the mind's infancy stage of development. The chimpanzee and other apes are still in mid-infancy where humanity was about eight million years ago. Love-indoctrination required ideal nursery conditions which means our description of the luxurious Rift Valley of Africa, where humanity spent its infancy and childhood, as the 'cradle of mankind' was particularly apt.

Succeeding at love-indoctrinating infants was a difficult task requiring all women's attention. With women so preoccupied nurturing, men had to support them. Men were subservient to the needs of women at this stage. The support men could give was to shelter and protect the group from outside threats such as marauding leopards.

What happened two million years ago was that a monster 'leopard' — a monster threat to the group — appeared. It was the threat of ignorance, ignorance on the part of our conscience (and, by association, all things innocent) of our need to attempt to understand existence and master self-management.

As group protectors it was men's role to meet this threat and, though they did not know it then, it would take two million years to vanquish. So great was this threat that it became humanity's dominant concern. If understanding could not be found, if our innocent conscience's efforts to stop our search for understanding could not be fought off, all future development of humanity and of the development of order on earth would be stalled. (This was the larger view. As has been explained the view from each man's position was that he had to try and realise his mind's potential or 'achieve as much as he could'.) At two million years ago, after ten million years of being a female role dominated or matriarchal society, the threat of ignorance turned humanity into a male role dominated or patriarchal society.

Not responsible for the fight to defy the ignorance of innocence, women's innocence became a victim of that fight. Women's innocence, like that of the animals, represented a defiance of the need to search for understanding, leaving men no choice but to either defy/oppress/destroy/attack that defiance or abandon the search. Unlike the animals, the destruction of women's innocence didn't entail destruction of women themselves. They couldn't be destroyed because they reproduced men. Instead sex, originally for procreation, became perverted, 'misused' by men as a means of attacking the innocence of women. This violation of women's innocence, this 'misuse' of them, this rape, made women as soul-destroyed as men but it brought them into harmony with men's desperate battle to defy ignorance. Women were adapted to the battle. Women came with men on the long march through humanity's adolescence, bringing them the only warmth and 'comfort' they would know. Civilised sex (as opposed to rape) became, in the grander sense, an act of love — an act of compassion, sympathy and support.

This oppression of women was a horrible situation. It needs to be explained more fully so it can be clearly understood. Not responsible for the fight against ignorance and so not partaking in the battle itself women could not know what that fight involved. Consequently if they were not oppressed they could naively impose their innocence. The effect of this would have been to deny humanity its need to be, to a degree, free to search for understanding. For example, if women had had their way, in their innocence, they no doubt would have stopped the men's 'bloodthirsty' 'sport' of hunting animals. Denied expression for their embattled egos — for their conscious thinking self or mind's need to, in some way or other, win against their innocent conscience's unjust criticisms of their efforts to master self-management — men could not have coped. The eight-year-old at the birthday table had to have some way of retaliating against the criticism from the innocents. He had to do something to defend himself. He could not be expected to just sit there and take it. If he were expected not to retaliate in any way then he could not be expected to be inclined ever to take the cake again, in which case he would never gain understanding. Men had to be allowed to express their egos and it can be seen that they would have to have extremely powerful egos if they were to have the necessary determination their job of championing the mind demanded. Their ego had to be strong enough to defy the innocence of the whole world because the whole world was an innocent friend of our soul, not a friend of our corrupt and apparently bad mind. But women did not and could not be expected to understand this battle. They could understand the search for truth but not the battle involved in that search. This extract from a newspaper article illustrates the problem: 'Shirley MacLaine can't find a man to love. 'The 48 year old actress [said] she longs for a "close and warm relationship", but hasn't met a suitable partner. "Most men I meet seem to be too involved in trying to be successful or making lots of money", she said. "I feel sorry for all of them. Men have been so brainwashed into thinking they have to be so outrageously successful — to be winners — that life is very difficult for them. And it's terribly destructive, as far as I am concerned, when you are trying to get a serious relationship going." '[1] What alternative was there but oppression? (It should be immediately stressed that this was another situation requiring balance. Too much freedom for women and they would stifle the search for understanding; too much oppression and they would become too dispirited and, as will shortly be explained, too soul-destroyed to effectively nurture a subsequent generation.)

---

[1] From Sydney's *Daily Mirror* newspaper, December 14, 1982.

The only alternative to oppression was that men explain themselves to women but the great tragedy was that this was not possible. Men could not admit their inconsistency with integrative meaning until they could defend it — least of all to an innocent who could only interpret a failed attempt to explain upset as an admission of 'badness'. Here is a quote from a magazine that makes the point: 'One of the reasons that men have been so quiet for the past two decades, as the feminist movement has blossomed, is that we [men] do not have the vocabulary or the concept to defend ourselves as men. We do not know how to define the virtues of being male, but virtues there are.'[1]

Feminists did not free themselves just because men stayed quiet as this quote suggests. The feminist movement was far more meaningful than that. The more men fought to defeat ignorance and protect the group (which was humanity) the more they became angry, egotistical, alienated and superficial and so the more they appeared to make the situation worse. The more men tried to protect us from danger the more they seemed to expose us to danger! In the end they became completely ineffective or inoperable, paralysed by this paradox. At this point women had to take over the day-to-day running of affairs and try to nurture a new generation of soundness. Women, not oppressed by the towering responsibility and extreme frustration that men felt, could remain effective. As well, when men crumpled women <u>had</u> to take over or the family, group or nation involved would perish. A return to matriarchy, such as we have recently been seeing on earth, was a sign that men in general had become completely exhausted. However it was not true matriarchy, because men could not afford to stand aside completely. They still had to stay in control of the fundamental battle, they still had to remain vigilant of the threat of ignorance. While some elements in the recent feminist movement, on sensing power coming women's way, took the opportunity to get even

---

[1] Article by Asa Baber in *Playboy* magazine, July 1983.

with men for men's oppression of them, the movement in general was most necessary and valid.

Throughout the battle to find understanding women were being forced to suffer the destruction of their soul, their innocence, at the same time having their trappings of innocence cultivated. Originally, as explained earlier, cute, youthful, childlike features were sought, were considered 'beautiful', because they indicated a potentially integrative person, they were the hallmarks of innocence. Later, when ignorance became a threat, men sought such 'beauty', such signs of innocence, for sexual destruction. We evasively described such looks as 'attractive'. It was an evasive description because we were avoiding saying that what was being attracted was destruction, through sex, of women's innocence. Because all other forms of innocence were being destroyed this cultivation of the beautiful object of innocence in women was the only way men were able to cultivate innocence during humanity's adolescence. Women's beauty became men's only equivalent for the beauty of their lost pure world. As the writer Albert Camus once said, 'women are all we [men] know of paradise on earth'. It was little wonder men fell in love with women. The 'mystery of women' was that it was only the physical image or object of love that men were falling in love with. For their part women were able to fall in love with the dream of their own 'perfection' — of their being truly innocent. Men and women <u>fell</u> in love. We abandoned the reality in favour dream. It was the one time in our life when we could be transported to 'how it could be' — to paradise.

This process of the destruction of women's souls and the cultivation of their image of beauty has been going on for two million years. Lust and the hope of falling in love have almost ruled our lives. So much so the famous psychoanalyst Freud was misled into believing sex did rule our lives. However in truth the battle to find understanding always remained. Men and women became highly adapted to their roles. While men's magazines are full of competitive battleground sport and business, women's magazines are full of ways for women to be 'attractive'.

Each generation of women had a very brief life (in innocence) before individually they became soul-destroyed through sex, after which they had to try to successfully nurture another generation into existence — all the time trying to conceal from that new generation the destruction that was all around the women and — after being soul-destroyed through sex — in them.

Women's tasks of having to inspire love when they were no longer love, were no longer innocent (how could they be having been dragged by men through the horrors of humanity's adolescence for two million years), to 'keep the ship afloat' when men crumpled and to attempt to wholesomely love another generation into existence — and all this with men dominating, and, as well, not being able to explain why they were dominating, what they were actually doing and why they were so upset and angry — was an altogether impossible job yet women did it and have done it for two million years! It was because of women's phenomenally courageous support that men, when civilised, were chivalrous towards them, holding doors open for them and giving them their seat in a bus. While men had an impossible fight on their hands at least they had the luxury of knowing what was going on.

The more men searched for and progressed towards understanding, the more angry, egocentric and alienated they became, the more they were upset by innocence, the more they were destructive of women's innocence, the more they became upset with themselves, the more they needed to find relief from that upset. (Interestingly the many paradoxes involved in this upset state of the human condition are perfectly described on an old poster for the movie *The Treasure of Sierra Madre*. The blurb reads 'Humphrey Bogart storms the terror-swept goldlands — a new high in high adventure. The nearer they got to their treasure [understanding], the farther they got from the [integrative] law! And the more they yearn for their women's arms [sexual release], the fiercer [the more upset they became with themselves and thus the more] they lust [and search] for the gold [liberating answers] that cursed them all.')

The point to be made here is that women's exhaustion development was tied to men's. Women had to try and 'sexually comfort' men but also preserve as much real innocence in themselves as they could for the nurturing of the next generation. Their situation, like men's, got worse at an ever-increasing rate. The more women 'comforted', the less innocence they retained, the more the next generation suffered, the more that following generation needed 'comforting', etc. If humanity's battle had gone on another few thousand years all women would have ended up like Marilyn Monroe, complete sacrifices to men, at which point men would have destroyed themselves and the species because there would have been no soundness left (in women) to love/nurture a fresh generation. Olive Schreiner in her novel, *The Story of an African Farm* (1883), made the same point. When talking of men persuading women to have sex she said (through her female character) men say ' "Go on; but when you [men] have made women what you wish, and her children inherit her culture, you will defeat yourself. Man will gradually become extinct . . . Fools!" she said!'

The convention of marriage was invented as one way of containing this spread of exhaustion. By confining sex to a life-long relationship, the souls of the couple could gradually make contact and be together in spite of the sexual destruction involved in their relationship. Brief relationships kept souls repressed and spread soul repression. Sex killed innocence. During humanity's adolescence that was what sex was all about, although it was also one of the greatest distractions and releases of frustration and, on a higher level, an expression of sympathy, compassion and support — an act of love. Once again we see the paradox of the human condition at work. By repressing sex and sexual attraction, such as the custom of all-over covering of women in Moslem societies, we could restrict the spread of sexual destruction but, as with humanity's exhaustion in general, not stop it completely. Only the arrival of the full truth or the answers could halt the development of exhaustion. For women to be genuinely liberated, men's upset had to be resolved. To do this the true defence for our mistakes had to be found — as it now is. All our upsets unravel from the source or original upset which was 'the taking of the cake' — humanity's effort to master self-management.

In Olive Schreiner's novel she goes on to say (again through her female character) that: ' "If I might but be one of those born in the future; then perhaps, to be born a woman will not be to be born branded." . . . "It is for love's sake yet more than for any other that we [women] look for that new time." She had leaned her head against the stones, and watched with her sad, soft eyes the retreating bird. "Then when that time comes," she said slowly, "when love is no more bought or sold, when it is not a means of making bread, when each woman's life is filled with earnest, independent labour, then love will come to her, a strange sudden sweetness breaking in upon her earnest work; not sought for, but found. Then, but not now. — " ' This future that women have dreamed of has arrived. Men's battle is won — is over. The source 'dragon' of all our dragons is slain. The 'devil' who/which, when we were innocent, was the coercion from the evasive world to abandon and thus compromise our ideals, or, when we were exhausted, was the false implication from the ideals that we were bad, is overcome. We now have the truth about ourselves. Our identity is found.

It should be pointed out here that the destruction of inno-cence, such as the destruction of animals and corruption of women just described, has been going on at all levels. Humans also destroyed the innocent soul in themselves by repressing it. All forms of innocence unfairly criticised humans, so all forms of innocence were attacked by us. The wearing of dark glasses ostensibly as sunshades was often an effort to alienate ourselves from the natural world that was alienating us — was an attack on the innocence of the daytime.

The paradox was that having destroyed innocence men would end up wanting to rediscover it. The truth was men were having to repress and 'hurt the one they loved'. To quote an exceptional modern day prophet or unevasive thinker, Sir Laurens van der Post, from his book, *The Lost World of the Kalahari* (1958), 'I thought finally that of all the nostalgias that haunt the human heart the greatest of them all, for me, is an everlasting longing to bring what is youngest home to what is oldest in us all.' While women's oppression has been extreme so has men's and men have yearned for freedom from their oppressor, ignorance, as much as women have yearned for freedom from their oppressor, men.

(Incidentally, it was impossible for a woman to be a prophet. Women could be exceptionally honest, often, in their naivety, more honest than men, as Olive Schreiner was, but not appreciating the threat of ignorance they were not in a position to reconcile the upset on earth because at its base the upset was caused by the threat of ignorance. For example Olive Schreiner called men 'fools' and Shirley MacLaine thought they must be 'brain washed'. Women certainly knew of upset but they did not understand it. Women were unaware of the threat of ignorance. The female equivalent of a male prophet was an exceptionally innocent mother capable of giving a son only pure love and thus producing a prophet. The primary role of women was nurturing.)

Having to assume the day-to-day running of affairs as men everywhere became exhausted made it difficult for women to give children the nurturing they needed. With men exhausted and women working this was becoming a serious problem for humanity because it would produce an even more exhausted generation. Now that the battle to get the truth up is won the whole situation can change and all importance be placed on nurturing a break-free generation.

Another dangerous distortion in nurturing was the tendency for parents when exhausted to transfer their ego needs onto their children and try to make them winners. Children are not competitive. Ego doesn't manifest until adolescence. Parents' efforts to 'jump-start toddlers in pre-school hot houses' and 'train them to be geniuses and super-kids'[1] was a sign of the arrival of the ego desperation that came with complete exhaustion. The more the mind searched for understanding the more it was criticised (by the conscience) and so the more it tried to win against the false implication that it was bad. It was our intellect or intelligence that was under siege — that was insecure. The more exhausted we became the more we tried to champion the intellect. In the end nothing but intellectual supremacy could be tolerated and we excessively repressed our soul. We became unbalanced.

Parents were not the only ones with excessively insecure egos or intellects. On a much larger scale our whole academic system placed excessive emphasis on the need for I.Q. in inquiry. Everywhere the need for more I.Q. was being stressed when what was required was more soul — was more soundness of self. To make the point, the <u>minimum</u> I.Q. requirement for entrance to any university in the world we like to name was a level that was too high for effective inquiry into the truth. How could this be? Once the boy took the cake at the birthday party and was criticised the only thing that would relieve the criticism was understanding. As soon as he took the cake the race was on to find understanding to stop the criticism, so suddenly there was a need for I.Q. or intelligence to find that understanding. Our mind or ego, or effectively each of us, became driven to find relief.

While the brain size of the australopithecines was not much bigger than Infantman (such as chimpanzees), there is a sudden increase in brain size in the first of the Homos, Homo habilis. That dramatic growth continues through Homo erectus and Homo sapiens finally to plateau off in Homo sapiens sapiens. Anthropologists have long wondered why this growth stopped. The reason is that in Homo sapiens sapiens a balance was struck between the need for cleverness and the need for soundness. The average I.Q. of people today is that amount which is safely conscience-subordinate. Too much I.Q. and we diverged too

---

[1] *Time* magazine, April 7 1986.

quickly and too far from our soul and all the ideals it knows; too little I.Q. and we were too conscience-obedient. We needed both the guidance of our conscience and our intellect's capacity for insight. While we recognised that insight and I.Q. were related we evasively failed to recognise that alienation and I.Q. were also related. The more intelligent we were, the sooner we took the cake, the sooner we repressed our soul and all the absolute truths (such as integrative meaning) along with it. With access to integrative meaning repressed it was impossible to think straight. Trying to make sense out of existence while avoiding integrative meaning was like trying to discover how a car worked having decided not to look under the bonnet at the engine. On the other hand too little I.Q. and we would never take the cake and never understand.

The average I.Q. of humans today was the ideal or most balanced I.Q. for inquiry into the truth, not the exceptionally high I.Q. that our academic institutions evasively sought. Ideally the exceptionally 'clever' should have been excluded from inquiry just as we excluded those who were exceptionally lacking in 'cleverness' or I.Q. Like the egotistical parents, academia was behaving extremely insecurely. Humanity had become dangerously over-exhausted. By excessively repressing soul/soundness/innocence in inquiry we were denying instead of cultivating the arrival of the liberating full truth. The answers would come from back down the road towards the world of soundness/innocence/soul. Intellectualism was a cul-de-sac in development.

A leading academic recently said that 'Biology has not made any real advance since Darwin'[1]. In Darwin's autobiography he says 'When I left school I was for my age neither high nor low in it; and I believe that I was considered by all my masters and by my father as a very ordinary boy, rather below the common standard in intellect.' It was mentioned earlier that trying to make sense out of existence while avoiding integrative meaning was

---

[1] Charles Birch, retired Challis Professor of Biology at Sydney University, in conversation with this author on March 20, 1987.

like trying to understand how a car works having decided not to look under the bonnet at the engine. The extent of the limitation of this blindness becomes abundantly clear when the reader sees just how much mystery has been cleared up in this book simply by living with instead of evading integrative meaning — simply by 'looking under the bonnet'.

It was in evading the truth that cleverness was necessary. It was our ability to evade what was really so obvious that was so brilliant. It is the story of *The Emperor With No Clothes* — where it took the sheer simplicity/innocence of a small boy to break the spell and expose the truth — and also the story of David and Goliath, where it took a small boy in all his simplicity and innocence to walk out and destroy the monster of our evasions which is Goliath. As Christ said (and this quote will be mentioned again very shortly) 'You have hidden these things from the wise and learned, and revealed them to little children'. The strength that was required to find the understandings in this book was the ability to defy our evasions, which, in the final analysis, is innocence or soundness, not cleverness. Through evasive, cleverness-stressing, mechanistic inquiry it was possible for humanity to find the pieces of the jigsaw of explanation but the pieces had to be presented upside down to hide the pictures of the hurtful partial truths they represented. (For instance, science discovered the law that explains development — The Second Path of the Second Law of Thermodynamics — but presented that hurtful partial truth evasively by stressing only its divisive direction towards entropy or disorder and calling the whole process aimless 'evolution' instead of purposeful development.) The ability to assemble these 'jigsaw pieces' to reveal the full truth required innocence because only innocence of hurt could look at the hurtful partial truths without being hurt. (For example, if you are not upset/divisive you don't have to evade integrative meaning.) The jigsaw could not be put together to reveal the full picture without looking at the pieces picture-side up. The full truth could only be found by confronting instead of evading the hurtful partial truths. The full truth could not be found using evasions/'lies'/ 'untruths'.

Humanity had to progress towards the truth evading any hurtful partial truths along the way. In the end this meant we had accumulated a mountain of evasion that only someone innocent and thus unevasive could dismantle. The army of humanity had made all the preparations it could. It had to await the appearance from among its ranks of an exceptional innocent — a David to go out and slay Goliath — to go out and overcome all the evasions in our evasively presented insights and reveal the full truth they contained and free us. But the problem was we were repressing innocence, denying its involvement.

By cultivating intellectualism we overshot the mark. We passed over our most effective thinkers. Sir Laurens van der Post has made the same point. He has been quoted as saying that some of the most anonymous people are among the greatest he has known, one of them being a Zulu who cannot read or write[1]. Antoine de Saint-Exupery (who wrote the *Little Prince* mentioned earlier) has been quoted as saying that the three greatest human beings he had ever met were three illiterates, two Brittany fishermen and a farmer in Savoy. He added, 'mistrust always the quick and brilliant mind'[2]. Christ warned us about intellectualism most determinedly almost 2,000 years ago but we did not heed him. He called the intellectuals of his day: 'You brood of vipers! . . . You blind guides' (Math 23) and said 'You [God] have hidden these things from the wise and learned, and revealed them to little children' (Math 11:25). He begins his most important speech, 'The Sermon on the Mountain', thus: 'Blessed are the poor in spirit'. Spirit was our original description for intellect. Even Einstein, whom we regard as an exceptional intellect, recognised the need for introspective guidance to accompany the intellect. He said 'Science without religion is

---

[1] From article by John Murche in *The Weekend Australian* newspaper, June 27, 1987.
[2] From *War Within and Without*, 1980, by Anne Lindbergh, page 30.

lame, religion without science is blind'[1] and in his obituary is enshrined a quote of his which says 'the cosmic religious experience is the strongest and the noblest driving force behind scientific research'. Introspection or holistic subjectivity with its requisite degree of soundness was as important in inquiry as research or mechanistic objectivity with its requisite degree of cleverness — the soul and the mind had equally important roles to play. We failed to heed the advice of even our intellectual heroes. We would listen to no one!

Actually it should be acknowledged here that science has been beginning to admit the dangers of being too mechanistic/evasive. Harvard scientist Stephen Jay Gould was mentioned earlier as one of a few scientists actively trying to resist and even expose the evasions of science. Another that could be mentioned is Charles Birch, former Professor of Biology at Sydney University in Australia. For instance Professor Birch has said:

'There are two ways of trying to understand nature. One is to reduce living organisms to next to nothing, such as atoms or their parts and then try to build up a world from these so-called building blocks. This is reductionism. . . . [It] is the dominant mode of science and is particularly applicable to biology as it is taught today. It leads to a materialistic or mechanical view of life which fails to do justice to what each one of us knows about being alive, namely, being creatures who feel and respond and have hopes, fears and purposes. A view or model of livingness that leaves out feelings and consciousness is an emasculated view of life. I believe it has grave consequences. . . .

In the name of scientific objectivity we have been given an emasculated vision of the world and all that is in it. The wave of anti-science and the profusion of cults and sects in our day is an extreme reaction to this malaise of materialism, mechanism, substance thinking or what you will.

I believe biologists and naturalists have a special responsibility to put another image before the world that does justice to the unity of

---

[1] From *Out of My Later Years*, 1950.

life and all its manifestations of experience — aesthetic, religious and moral as well as intellectual and rational.'[1]

Nobel Laureate biologist Jacques Monad in his book, *Chance and Necessity* (1970), gave a similar warning when he said:

'In the course of three centuries, science, founded upon the postulate of objectivity, has won its place in society — in men's practice, but not in their hearts . . . the choice of scientific practice . . . has launched the evolution of culture on a one-way path; on to a track which nineteenth-century scientism saw leading infallibly on to a vast blossoming for mankind, whereas what we see before us today is an abyss of darkness.'

It was no wonder the creationist alternative became established to counter science's overemphasis on Godlessness or denial of integrativeness. We had to come up with something to counter science's entrenched blindness. The only way we have had to maintain and preserve the absolute truths was to enshrine them as metaphysical concepts within our various old religions. By bringing science further and further to the fore and making these metaphysical truths increasingly remote and not of our world we left no effective presence of the absolute truths in our midst. Evasion was all dominating. In earlier times we had more respect for religion, which kept the door to the absolute truths open. Religions were the custodians of the truth while academia was the custodian of evasive insights into the truth. We needed a balanced presence of both but this was not maintained. Religions may have been remote with their mystical explanations but science was equally as remote and mad with its evasive explanations and this equality of craziness was not reflected by a balanced presentation of the two views. Science in its ego came to portray itself as being totally rigorous, responsible and sound when it was nothing of the sort. Science was

---

[1] From the *Australian Natural History* magazine, Volume 21 Number 2, 1983.

choked up with a mountain of evasion — a 'pack of lies'. It was extremely false.

The following quote illustrates what has been said here about the rise of creationist explanation to balance Godless science. The author, Dr. Gish, is the associate director of The Institute of Creation Research in San Diego. When he said the following he was advocating teaching Creationism in school:

'What has happened in our society in the last half century or so is that our young people in the colleges, universities and schools have been taught the theory of evolution as an established fact.

They've been taught that evolution is an exclusively naturalistic theory, and that God is not necessary. God, by definition, is excluded from the process. When the student hears this, he thinks we start with hydrogen gas and our only destiny is a pile of dust. Therefore there is no one to whom he is responsible. The teaching of the theory of evolution had caused the moral deterioration of modern society Today we have a rampant drug culture, legalised pornography, and abortion . . . now, we might ask ourselves why have these changes occurred.'[1]

The part that says 'our only destiny is a pile of dust' is a reference to the evasive emphasis of science on entropy, which implies disintegration is our destiny or meaning. Blaming science for all our woes was unfair although we can see in retrospect that Dr. Gish was correct in making his point strongly. Charles Darwin, in his book commonly referred to as the *Voyage of the Beagle* (1839), expressed a similar concern when he said: 'If the misery of our poor be caused not by the laws of nature, but by our institutions, great is our sin.'[2]

Professor Birch has said: 'Scientists tend to detach from the bigger view of things but science won't survive until science develops a stronger conscience'[3].

---

[1] From the *Sydney Morning Herald* newspaper, January 8, 1986.
[2] Taken from the half-title page of Stephen Jay Gould's book, *The Mismeasure of Man*, 1981.
[3] From the *Sydney Morning Herald* newspaper, November 12, 1983.

The problem was that to develop a stronger conscience — to stress soundness without defence for our lack of soundness — unfairly added to our sense of guilt. The truth is humanity in its insecurity was capable of recognising every talent except soundness. Christ said, 'foxes have holes and birds of the air nests but the Son of Man [those made 'in the image of God' — those unseparated from their soul or instinctive self — those unexhausted and thus capable of being integrative and thinking integratively — the exceptionally sound] has no place to lay his head' (Math 8:20). We did not want to know about the sound or soundness. The world of the sound, the way they thought, what they could see and the way they felt, went unrecognised on earth. The sound among us, like the soundness (our soul) within us, were ignored and repressed into anonymity. They were alone with their soundness, unwanted. We can appreciate this when we realise how repressed our soul itself has been. After two million years of repression our soul was as lonely within us and as cast out as a lost bird in an empty desert or sea. To quote Samuel Taylor Coleridge describing our 'soul in agony' in his poem *The Ancient Mariner*:

'Alone, alone, all alone
Alone on a wide, wide sea!'

When this book was about to be typeset for printing, the author's mother found the following passage which just has to be included even though it is rather lengthy because is is so expressive of all that has just been said. It appears at the start of the introduction to a book titled, *Simone Weil, An Anthology* (1986), edited and introduced by Sian Miles. It reads:

'Simone Weil completed her last work in 1943 when at the age of thirty-four she died in an English sanatorium. Two weeks earlier she had written to her parents:
"When I saw [Shakespeare's] Lear here, I asked myself how it was possible that the unbearably tragic character of these fools had not been obvious long ago to everyone, including myself. The tragedy is not the sentimental one it is sometimes thought to be; it is this:

There is a class of people in this world who have fallen into the lowest degree of humiliation, far below beggary, and who are deprived not only of all social consideration but also, in everybody's opinion, of the specific human dignity, reason itself — and these are the only people who, in fact, are able to tell the truth. All the others lie.

In Lear it is striking. Even Kent and Cordelia attenuate, mitigate, soften, and veil the truth; and unless they are forced to choose between telling it and telling a downright lie, they manoeuvre to evade it. . . . Darling M., do you feel the affinity, the essential analogy between these fools and me — in spite of the Ecole and the examination successes and the eulogies of my 'intelligence' . . . [which] are positively intended to evade the question: Is what she says true?' And my reputation for 'intelligence' is practically equivalent to the label of 'fool' for those fools. How much I would prefer their label."

Since then she has become known as one of the foremost thinkers of modern times, a writer of extraordinary lucidity and a woman of outstanding moral courage.'

Our soul had no friends amongst humans today. All it had was the knowledge of, and enthusiasm for the utterly beautiful world where it came from which, in spite of its repression, represented a more powerful force than any other force on earth — nothing man has created can remotely compare with the beauty and happiness of our lost world. In recognition of this power the word enthusiasm is derived from the Greek word 'enthios' which means 'God within'. The problem in the first place was that our soul was unfriendly towards our mind. It unjustly criticised our necessary efforts to master self-management which meant we had to be tough enough to repress it or fail to fulfil our responsibility to learn self-management. Our soul cast out our mind which left our mind no choice but to retaliate and cast out our soul. As an expression of our soul the sound also represented unjust criticism of us. Soundness reflected critically on our lack of soundness — a position we were unable to defend. The full truth is that those of us who were sons of upset were just as much sons of God as those of us who were sons of righteousness or the image of God or integrative or sound. We could not acknowledge the integrativeness/Godliness of the sound until we could explain that the exhausted were also Godly. We could not recognise, cultivate or institutionalise soundness until we could defend exhaustion. We could not recognise our soul until we could defend our mind. The existence of soundness/our soul was a hurtful partial truth that we had to evade. As has been mentioned we cultivated a mechanistic/objective/evasive approach to inquiry. We could not cultivate a holistic/subjective/unevasive approach. Like the end of the oppression of women, the end of the oppression of soundness — of our soul, of our original beautiful world — depended on the end of the oppression of our intellect by ignorance.

Prophets, women, animals, nature and all other forms of innocence had to suffer repression and persecution. There was no other way. The intellect/spirit had to succeed, had to achieve its goal. However excessive repression of women and prophets — of the ability to nurture innocence and of innocence or soundness itself — and the intellect would actually fail to achieve its liberation from criticism. Excessive exhaustion, particularly alienation and superficiality, and we would never find the truth. Such a situation of extreme 'lostness' has been occurring on earth. For example: 'according to a new book, *Women and World Religions*, oppression of women . . . is due to womb envy on the part of men'[1]! Any more superficial and we would take seriously the Monty Python answer that 'the meaning of life is the number 42'. Our planet has been awash with superficial answers. The word superficial means the opposite of deep or profound. The desire not to 'dig deep' is what alienation is. So the earth was awash with alienation. Now that alienation can be defended the sound can be acknowledged for their soundness. The priority now that we are at last defended is to nurture soundness and to bring the sound to the forefront of our development to clear up

---

[1] From the front page of the *Sydney Morning Herald* newspaper, July 2, 1987.

any remaining evasion, alienation and superficiality and lead us out of our development cul-de-sac.

To free ourselves from our embattled state we had to become secure in our understanding that our intellect and the exhaustion that it gave rise to was not bad — that all humans are equally good. It can be understood now that we all fought equally hard for humanity with the different stages of exhaustion from fighting accounting for the differences in our behaviour. In truth the exceptionally sound were not 'great' as van der Post and Saint-Exupery claimed, only exceptionally innocent and therefore capable of being exceptionally sound and unevasive. They were just new to the battle. If a fresh player runs onto the football field three-quarters of the way through a game and makes some hard and straight runs with the ball it doesn't mean he is a better player than the others, only that he is less exhausted. Humans are differently exhausted but no human is superior or inferior to another. We were all equally good soldiers for integrativeness — for God — with the only difference being that some of us had been fighting for God longer than others — some of us have been more heroic than others — some of us have been searching for understanding and battling the accompanying unfair criticism from our soul longer than others.

Having been through a period of intellectualism humanity will now go through a period of 'soulism' to clear up any remaining evasions. The situation now is that the exhausted can go into therapy (instead of into even more egotistical, futile, false and upsetting pursuits as has often been the case) since therapy is now at last possible, and the sound can go into inquiry since they no longer represent criticism of the rest of us. Effectively organised at last (after all, the practical way to play a game of football is to have the fresh players on the field and the exhausted ones off the field resting), we will quickly mend ourselves and our earth. When all our evasions are cleared up there will be no further need for our soul. The intellect will then be free to realise its full potential as the master tool that it is. It is the nerve-based learning system, or mind, that can knowingly integrate the universe. Also, while humanity will now start to close its cities down (in truth cities were not functional centres as we evasively claimed, they were hideouts for alienation and places that perpetuated/bred alienation — to quote the Australian historian Manning Clark 'the bush [wilderness] is our source of innocence; the town is where the devil prowls around'[1]) and go back to nature to rehabilitate its soul and become sound again, it shouldn't 'throw out the baby with the bathwater' and lose all the knowledge it gained in inquiry such as in the field of high technology. While our energies were often horribly misdirected and our creations often as extremely distorted as we were, nevertheless, in our mad and driven state, we did cover a lot of valuable ground in inquiry that will be needed when we set about integrating the universe.

This time the story has got a long way ahead of itself. In progressing through the stages of adolescence we were up to the so-called 'hunter-gatherer' lifestyle that was perfected by Homo erectus. It was during the 'hunter-gatherer' existence that we refined all our conventions for coping with the human condition that we now take for granted, such as marriage to contain sexual destruction. We settled into the long and painful journey to find understanding, although at the end were racing against the onset of total exhaustion. As an indication of the increasing speed at which exhaustion was developing, Adventurousman, Homo erectus, existed for a million years but Angryman, who was Homo sapiens, lasted only half as long, from half a million years ago to fifty thousand years ago — before maturing into Sophisticated (in exhaustion) man who is us, Homo sapiens sapiens, and now, only fifty thousand years later, Sophisticatedman is giving rise to Triumphantman, or Godman. The emergence of Triumphantman signals the end of humanity's search for its identity, which was its adolescence, and the arrival of humanity's adulthood where it has to implement that identity and knowingly manage the development of order of matter.

---

[1] From the *Sydney Morning Herald* newspaper, February 18, 1985.

The hardships and confinement of life during the ice ages contributed to the speeding up of this progression because these periods dramatically accentuated the difficulties encountered by humans co-existing under the strain of the human condition. As has been mentioned before, isolation from encounter with the battle minimised the spread of exhaustion. If we were each alone with our level of exhaustion we would not be criticised by the fresher souls or corrupted by the more battle-worn. It was because of this truth that we often said we 'had to make an effort' if we were to go out and be social. The closer humans lived together during humanity's adolescence and/or the more difficult the living conditions the greater the occurrence and spread and thus increase in upset. An ice age represented one very long trying winter. In fact out of each of the great ice ages the next more exhausted stage of man appeared.

To return to our twenty-one-year-old (Homo erectus on the anthropological scale) who went out into the world to learn — to apply himself — to 'achieve as much as he could'. He gets a job and starts to master it, so beginning to play his part in humanity's quest for understanding. Gradually 'life's compromises' (the compromises he has to make to his soul) change him from an idealist to a realist. Gradually he comes to experience to the full extent the difficulty of practising self-management without an adequate defence for the mistakes that result. Through experience he learns sympathy for the highly imperfect real world. Because of the compounding effect of upset he becomes increasingly embattled and, in the end, desperate. From his position of sympathy towards the real world he becomes an outright supporter, attacking anything 'ideal'. By his late thirties he is in a rage of anger and viciously determined to win against the unfair criticism he is experiencing. He becomes totally embattled or 'punch drunk'. On reaching this state of absolute hate and destructiveness he begins to hate even himself.

Still lacking the exonerating answers that could relieve his anger, all he can do is learn to discipline himself, contain his rage. Through bitter experience he learns to rein in his expressions of upset. Nearing forty years of age, he learns to civilise his upset. When humanity became civilised it hadn't eliminated its upset only learnt to contain it. Civility didn't solve upset, it only disguised it. The anger was still there, only repressed and restrained. Instead of expressing what he feels our forty-year-old learns to ask acquaintances politely how they are faring and talk about harmless things such as the weather.

Such civilities, while they made living together possible, were an extreme form of pretence — of being what we were not. This falseness, while highly destructive to any young innocents looking on, was far less so than expressing the real upset. We had no choice but to become sophisticated (in evasion and repression). Angryman, Homo sapiens, lived out the upsets that were mostly repressed into only fantasies in us, his descendant, Sophisticatedman (Homo sapiens sapiens). For example it was during the life of Homo sapiens that we needed to 'lock up our daughters' as the saying goes, for he lived out the sexual destructiveness that we, Homo sapiens sapiens, have long since learnt to repress, to civilise.

All these stages of exhaustion during a mental lifetime are now partially instinctive in us. We have become adapted to the battle. For example, of 100 men trying to restrain their anger we could expect a few to have a nature (a genetic make-up) that in some way would make it easier for them to do so, so they would make the adjustment and cope and thus survive better. In each generation, this would have been the case, so the genes, as it were, followed the mind, 'reinforcing' the adjustments the mind was making. In this way all the adjustments described have been partially genetically built into us and, to a degree, appear automatically as we reach the ages where they are needed. Being so pre-adapted to cope we have often hardly realised the psychological adjustments we have been making as we grew (actually died in both soul and spirit) — except obliquely when we talked about generation gaps and acknowledged the wildness of youth, etc. These stages of adjustment account for the way people of particular ages have been only really able to find empathy with

74

others of almost their exact age. A lot of awareness and, with it, knowledge has been age-locked.

Not only were there different stages with ages, there were also different instinctive types of lives to cope with whatever degree of exhaustion we might have to adjust to when we were born. The stages with ages being described are the average mental life-time, the main one that we are adapted to. However, we could be born into the midst of the battle and have to start our life from a position already well down the exhaustion curve. The offspring of the boy who took the cake will grow up with a father angry, egocentric, alienated and superficial. Since this is not the sort of loving behaviour our soul expects and unable to be told why this 'mistreatment' is occurring the souls of the offspring will become upset also. In this way the 'sins' (the critical biblical description we had for our upset behaviours) 'of the father' will be passed on from generation to generation (see Ex 20:5). Or we could be born into a sheltered recuperative corner of the battlefield and start our life from a position of exceptional innocence. Inevitably in the thousands of generations that led to us every variety of life possible under the human condition would have been encountered and the fact that our ancestors survived to produce us means they must have had what it took to cope, therefore we can expect to be similarly adapted. We are the product of a great deal of genetic refinement. The subtlety this introduces is that while the type of life we would live would largely be chosen by the degree of 'nurture' in our upbringing there was a lot of 'nature' involved in that life. Through generations of experience, life at every level of exhaustion has refined its own ways of coping and maximising its position. It was as if there were a basket of different lives within us ready for use. Depending on what level of exhaustion we encountered when we were young — what rung on the exhaustion ladder we stepped off from or started our life from — an appropriate life would be drawn out for use. As occurred with different stages with ages so people of similar types of lives were better able to 'identify' (understand and thus sympathise) with each other. But these kinds of subtleties have to be explored in the full version of this book, not here in the condensed version.

Like the alienation within ourselves the alienation between different generations, ages and lives will be ended now that exhaustion can be explained and defended. We can begin to truly communicate with each other. To date, our wonderful forms of communication have often served only to communicate or spread exhaustion. Now they can be used to heal exhaustion. For example, we can now begin to talk about what is really going on inside ourselves instead of having to talk about our latest pair of 'attractive' blue shoes or our latest business 'takeover' or just the weather. Able to defend and explain ourselves to each other we will be able to understand each other. Everyone will be able to understand — to identify with everyone else. In truth we have each been terribly alone within ourselves. It is why we have often identified with such lonely places as the sea and the desert. At the other extreme, our inability to identify with each other — our alienation from each other, our insecurity about the fact of our goodness — was the basis of racism.

The author Germaine Greer once made the comment that 'growing like ageing went in leaps and bounds'. We will discover that a good deal of our ageing, like our state of health, was tied to our exhaustion development — was psychosomatic. So growing with, or adapting to the human condition and ageing to some extent have been related. Because of our now instinctive preconditioning we did not notice that we were advancing down the ever-steepening (deteriorating) exhaustion curve in leaps and bounds. Only occasionally when we were able to look back, such as happened when we heard music or smelt smells associated with our youth, did we get a glimpse of just how much deterioration had taken place. Nostalgia is defined as 'yearning for what is past or inaccessible; sentimental evocation of past happinesss'.

However, to return to our middle-aged, civilised man. Civilising his upset can only slow its increase, not stop it or reverse it. So he becomes even more lost from the ideal world. It is at this stage, as he enters his early forties, that he discovers religion (where we preserved the absolute truth of integrativeness) and is able to be born again into the world of our soul. By adhering to the absolute truths enshrined in his religion he is able to once again be an effective force for integrativeness in spite of his divisive, embattled state. He is reborn into effectiveness although his grasp of the ideals often reveals a strange unfamiliarity with them. For instance in his 'born-again' idealism he decides to support the remote 'save the seals' campaign, when, were he really sound, he would grapple with the much closer to home real problem — the non-ideal human condition. He has to satisfy his need to be ideal without confronting his exhaustion or, by association, humanity's real exhaustion. (As was explained earlier to be 'born-again' depended on abandoning or escaping our embattled state not on confronting it.) He is evasively unevasive. It is in many ways symbolic idealism. The unevasive, genuinely ideal way to end the devastation on earth was to confront our evasions and get to the full truth and thus fix the source of our upset — was to get the truth up — was to defy evasion not add to it. The problem was within ourselves not in the Arctic. Our born-again idealist overlooks the problem in himself, in his home and in his city where there is virtually no ideal/soundness/nature — or seals — left at all! In his evasion he 'strains out a gnat but swallows a camel' (Math 23:24).

Given the hope many people have been placing in causes such as the so-called 'Green Movement' to repair our earth the point made in this last paragraph needs to be emphasised. While we have had to evade the fact, the devastation of our planet and more importantly the deprivation and suffering of its people was entirely due to the human condition of upset — was due to our egomania, aggression and superficiality/blindness/alienation. Study any example closely enough, mental illness, famine, war, nuclear proliferation, corruption, air pollution, soil erosion, rainforest or wildlife destruction, to name a few, and we will see that this is true. Trying to solve these problems in any way that did not address the fundamental underlying problem of our upset was an evasion of the real problem and ultimately could not succeed. It was inevitable that our earth would end up as exhausted as we were. We could delay the debilitating process but not stop it. Only the finding of <u>the</u> answers/understandings that appease our upset could stop the devastation of ourselves and our planet. Our remaining wilderness areas are going to be vital for the rehabilitation of our souls but to save them we have to tackle the problem of our upset human state or condition. Believing that we could repair the earth without confronting our upsets, that we could set about developing a so-called 'steady state', or 'ecologically sustainable society', as Herman Daly, Professor of Economics at Louisiana State University, proposed in his book, *Steady State Economics*, was another evasion of the real problem. Believing that we could learn to discipline and contain our human greed, indifference, anger, aggression, insensitivity and destructiveness without tackling the source of these upsets was a giant self-deception, was a falseness that contributed further falseness. This truth can be admitted now, our false hopes and beliefs abandoned and the real problem of our upset addressed. The unevasive thinker or prophet Carl Jung, in his book, *Modern Man in Search of a Soul* (published many years ago now in 1933), was making the same point when he said: 'It is becoming more and more obvious that it is not starvation, not microbes, not cancer, but man himself who is mankind's greatest danger; because he has no adequate protection against psychic epidemics, which are infinitely more devastating in their effect than the greatest natural catastrophes.'

It should be mentioned that along with the exceptionally exhausted born-again idealists, such remote causes as the 'save the seals' campaign also attracted the exceptionally innocent and the exceptionally unaware who did not know of the real problem of the human condition. The realm of idealism was populated by both the exceptionally innocent and the exceptionally exhausted and because of the need that has existed until now to be evasive they were left undifferentiated.

To return to our born-again idealist he now also smiles all the time because he is now 'good' and no longer hurt but if he was truly unhurt he would be worried about the world.

Strange as born-again idealism has often been, the born-again idealist had a most important role to play. In an utterly exhausted world where true soundness had almost all been spent those 'born-again' to the ideal world were virtually the only source of integrativeness left to balance exhaustion's divisiveness. To be born again was highly responsible behaviour. Those among us who were irresponsible and lacking the necessary courage were those who held onto their angry, divisive selves and refused to be born again when they eventually became utterly spent.

This phenomenon of abandoning our embattled state and its world and embracing the ideal world has been expressed in different ways throughout our history. One of the most recent has been the adoption of the ideal future, the so-called New Age movement. It is important to realise that this movement, like the ones before it, practised artificial 'transformation', providing the same refuge traditional religions did, and was not a real transformation. As has been explained, a real transformation depended on confronting our blocks or alienations — depended on being unevasive — depended on finding understanding — depended on being able to think our split selves back together. Our real freedom lay back through our blocks not through abandoning them and distancing ourselves from them. New Age magazines and books typically chose 'positive' image cover illustrations such as a smiling 'healthy' girl back-lit with sunlight shining through leaves or nature-sympathetic pictures of rainbows. These images, like the contents of the books, were promoting a false freedom. A confrontation with our upset selves would look at the truth of our upset not evade it — would be illustrated by what the New Agers would term 'negative' image pictures such as artist Francis Bacon's horrific but truthful portrayals of the human condition. What the New Agers saw as 'positive' was actually a 'negative' in terms of achieving liberation from our upsets. In the main, the New Age movement was concerned with escaping to the future without traversing the path that takes us there. In the words of Jacob Bronowski from his book, *The Ascent of Man* (1973), 'I am infinitely saddened to find myself suddenly surrounded in the west by a sense of terrible loss of nerve, a retreat from knowledge into — into what? Into Zen Buddhism; into falsely profound questions about, Are we not really just animals at bottom; into extra-sensory perception and mystery. They do not lie along the line of what we are now able to know if we devote ourselves to it: an understanding of man himself.'

The New Age movement was advocating a false freedom and was mostly being led by false prophets or alienated/'blind guides' (Math 23:24). Exhaustion/alienation could not investigate alienation — could not reveal the truth. If it could it would not be alienated. Only innocence could investigate and liberate us from alienation. Ultimately it was necessary for the exhausted to admit their state and cultivate innocence, not deny their exhaustion and in so doing further deceive and thus repress innocence which added to our psychosis/evasion/denial. Now that we have the defence for our condition we can at last do this. We can afford to end all our deceptions/falsehoods. We can abandon all the false props and false ways of coping that sustained us this far. We can go into true therapy instead of into pseudo therapy and further escape.

We have to distinguish between true prophets and false prophets — between the sound and the born again. The two have constantly been lumped together with the latter trying to imitate the former although the truth is they are from the opposite ends of the exhaustion or departure curve. Because of this confusion it has been easy to discredit a true prophet by implying he is a mystic and concerned with the occult — by implying he is strange or psychotic or over-exhausted instead of sound. The aspect of the deception or disguise of their exhaustion aside, it was true, as stressed above, that the strange idealism/honesty of the born again did represent an extremely valuable source of idealism/honesty in a world almost devoid of it.

The New Age movement was a sign of desperation — of desperate need for relief — of a world that had become over-exhausted. Compared to the pseudo confrontations with our exhaustions being put forward by the New Agers — such as so-called 'channelling' (where 'mediums' supposedly talked with people from the past) and a 'wheat grass juice diet' and 're-birthing immersion therapy' and 'miraculous self-esteem achieved through reading and thinking mantras of positive thoughts', and 'healing the earth by holding hands and humming' and 'Zen-like [evasive] esotericism' — the real truth/idealism is unmistakeable. With our evasions at last confronted, torn down and replaced with the compassionate but nevertheless shocking full truth, as they now are in this book, it is only a matter of time before we come to recognise it. The full truth comes like rolling thunder. We have known this for a long time. For instance it is perfectly described, albeit somewhat emotionally (as is the character of introspection or subjectively found insights that have had to fight their way up through all our evasions), in both the Old and New Testaments of the Bible. For instance, see Mathew 24:24-35, which in part says: 'false prophets will appear and perform great signs and miracles to deceive even the elect. See I have told you ahead of time. So if anyone tells you, "there he is, out in the desert," do not go out, or, "Here he is, in the inner rooms," do not believe it. For as the lightning comes from the east and flashes to the west, so will be the coming of the Son of Man [the unevasive truth].' The reason traditional religions with their emphasis on guilt/badness, confession/apology and the need to ask for forgiveness have become unpopular is that we had become over-exhausted and desperate for relief from criticism. We have become completely embattled, incapable of accepting or admitting any negatives about ourselves. While the so-called 'humanism' of the more recent movements was legitimate (because humans were not fundamentally guilty), they could develop a lack of discipline and honesty that less-exhaustion-adapted traditional religions, such as Christianity, maintained with their emphasis on acknowledging our embattled state.

It should be added here that trying to escape to the future and in general crying out for it, even demanding it as the New Age movement was doing, was not of itself going to produce the New Age. While such desperation sought innocents/prophets it was not conducive to producing them. The nurturing of innocence required an environment that was free of battle-exhaustion, desperation and insecure superstitious nonsense. In truth the New Age was another declaration of hope. In the Sixties it was called the Age of Aquarius. Each age had its own ego — its own need to be fresh and unique — its own need to 'reinvent the wheel'. The New Age and the Age of Aquarius are new terms for a two-million-year-old hope, a hope described in the Bible as the dream of heaven. Similarly 'future shock' is a new term for 'judgement day' and 'collective unconscious' a new term for 'soul' and 'lateral thinking' is a re-description of 'imagination'.

However, to return to our story. Following his early-forties religious period our soldier enters the post-battle stage of his life. During his twenties he fought the battle of idealism where he tried desperately to hang onto the ideal world, during his thirties he fought the battle of realism, to defend the imperfect real world. In his forties he finally arrives at an overview of the whole journey. Through experience he finds, if not understanding, since that is dependent on still-to-be-found insights, then an awareness of the truth. He arrives at that state of compassion of sorts we term wise. He still cannot clearly explain it all but at least and at last he does know — does understand the human condition.

As a way of summing up the life of a human during humanity's adolescence it should now be possible to interpret our music's description of it. If we listen to a Beethoven symphony we can hear how lonely our soul is within us and how muffled yet still determined our will or spirit is. In young people's music we can hear the relative innocence, optimism and defiance of youth. They still have plenty of fight and resilience, plenty of 'rock and roll'. In more mature music like Beethoven's that strength is no longer so relentless but comes in surges of optimism and conviction. Above all, classical music reveals a compassion, expressing the fact that humanity's journey away from itself through the wilderness was something necessary and beautiful. This is a truth which stood above the terrible suffering, pain and atrocities involved.

The composer, Andrew Lloyd Webber, recently wrote a requiem inspired, he said, by the story of a Cambodian boy who had been forced by soldiers of Pol Pot to kill his sister. To rise above that sort of pain and still love the world has been the predicament of humanity. We have been through so much horror and yet the greater truth remains that we are still sublimely beautiful beings. How much agony must we humans have been in to be capable of that degree of brutality. We have been terribly undefended and unloved on earth.

In such rare instances of compassion, as sometimes appeared in our music, we have been able to find great truth, peace and serenity. In our music we could hear it all. We have not been able to talk in plain truth but we have been able to talk indirectly, and music is an excellent evasive language, nothing being admitted out of place. Our music said that with monstrous courage humanity was going to win the world's fight (to find understanding and reconcile our spirit with our soul) — and it has.

The Story of the Development of Humanity

(Portrayed as a journey via passes through mountain ranges that bar the way to becoming integrated or Godly.)

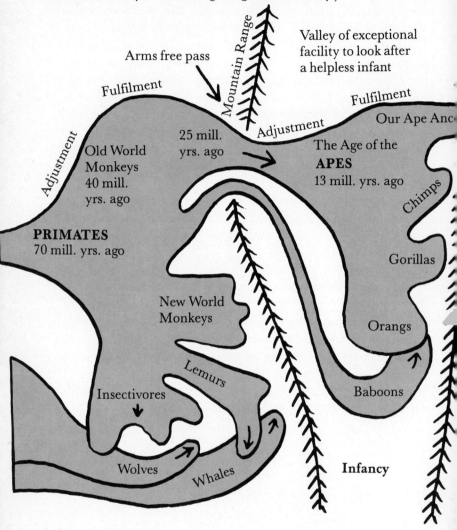

Note: *'Adjustment'* refers to the stage in development when adjustments occur in response to an opportunity to develop. Major changes to organisms take place in this stage.

'Fulfilment' refers to the fulfilment of the above adjustments. In this stage minor differentiations of a species can occur as different niches in the new realm and/or situation are occupied.

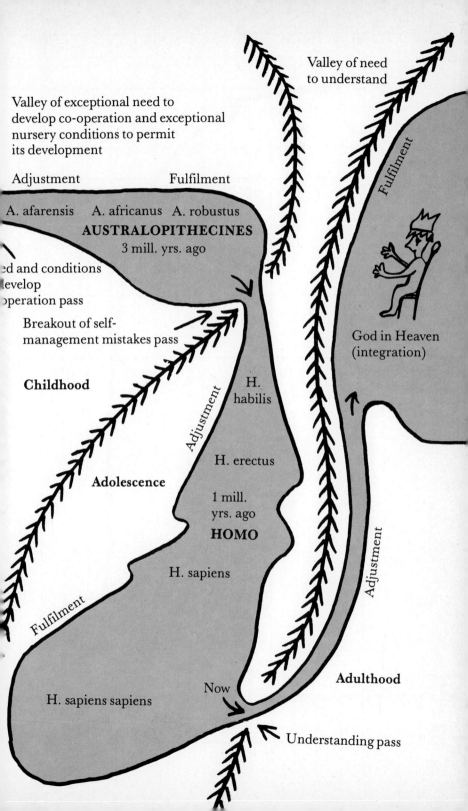

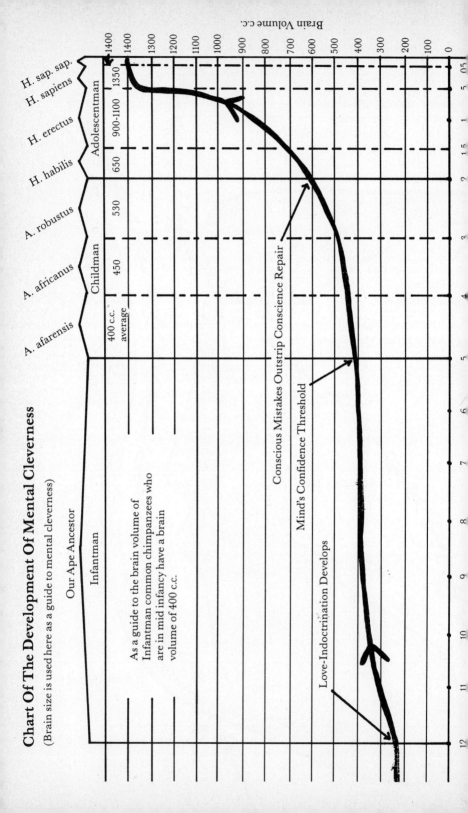

Chart Of The Development Of Mental Cleverness
(Brain size is used here as a guide to mental cleverness)

Brain Volume c.c.

	Brain Volume
H. sap. sap.	1400
H. sapiens	1350
H. erectus	900–1100
H. habilis	650
A. robustus	530
A. africanus	450
A. afarensis	400 c.c. average

Adolescentman

Childman

Our Ape Ancestor

Infantman

As a guide to the brain volume of Infantman common chimpanzees who are in mid infancy have a brain volume of 400 c.c.

Conscious Mistakes Outstrip Conscience Repair

Mind's Confidence Threshold

Love-Indoctrination Develops

Chart Of The Development of Integration

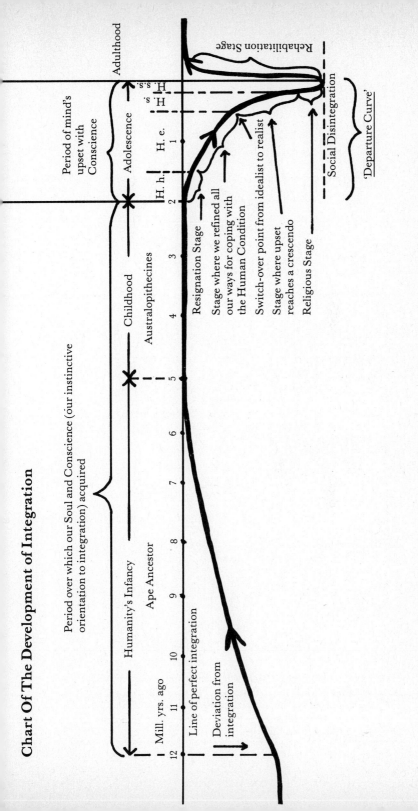

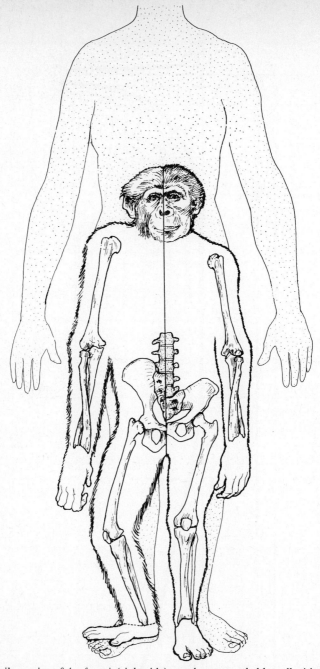

Fossil remains of *A. afarensis* (right side) match up remarkably well with the bones of a pygmy chimpanzee (left side). The main differences are that *A. afarensis* has larger teeth and a bipedal rather than quadrupedal pelvis. Size comparison is made with modern man *H. sapiens sapiens* outlined at rear. Drawing by Adrienne L. Zihlman.

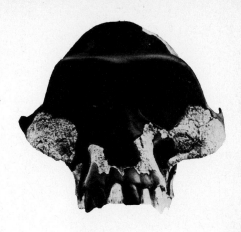

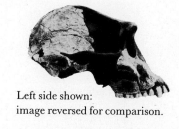

Left side shown:
image reversed for comparison.

(Images reversed)

**Early Prime of
Innocence Childman** ⟶

Australopithecus afarensis

The above photograph is of a
common chimpanzee skull (top),
an *A. afarensis* jaw (middle) and a
H. sapiens sapiens jaw (bottom).
(Photograph by John Reader ©
1981.)

Designation: Composite
Geologic age: 3 to 3.6 million years
Sex: Adult male
Discovery: M. Bush, 1975
Site: Hadar, Ethiopia
Housed: National Museum of
    Ethiopia, Addis Ababa

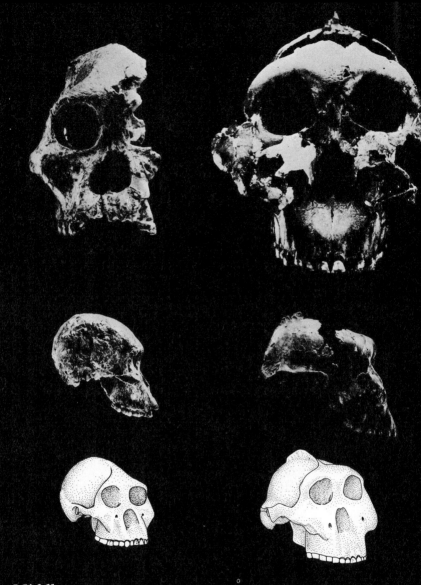

Middle Demonstrative Childman ⟶   **Late Naughty Childman** ⟶

Australopithecus africanus

Designation: Sts 71
Geological age: 2.5 to 3 million years
Sex: Adult male
Discovery: R. Broom and J. T. Robinson, 1947
Site: Sterkfontein, South Africa
Housed: Transvaal Museum, Pretoria

Australopithecus boisei

Designation: OH 5 ("Zinj")
Geological age: ca 1.8 million years
Sex: Adult male
Discovery: M. D. Leakey, 1959
Site: Olduvai Gorge, Tanzania
Housed: National Museum of Tanzania, Dar es Salaam

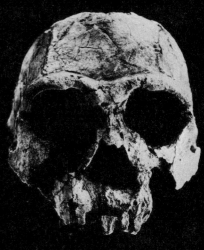
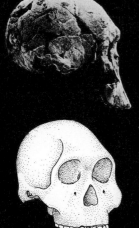

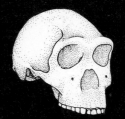

(Image reversed)

Sobered Adolescentman ⟶

Homo habilis

Designation: KNM-ER 1470
Geological age: ca 2 million years
Sex: Adult male
Discovery: B. Ngeneo, 1972
Site: Koobi Fora, Kenya
Housed: National Museums of Kenya, Nairobi

Adventurous Adolescentman ⟶

Homo erectus

Designation: KNM-ER 3733
Geological age: ca 1.5 million years
Sex: Undetermined
Discovery: B. Ngeneo, 1975
Site: Koobi Fora, Kenya
Housed: National Museums of Kenya, Nairobi

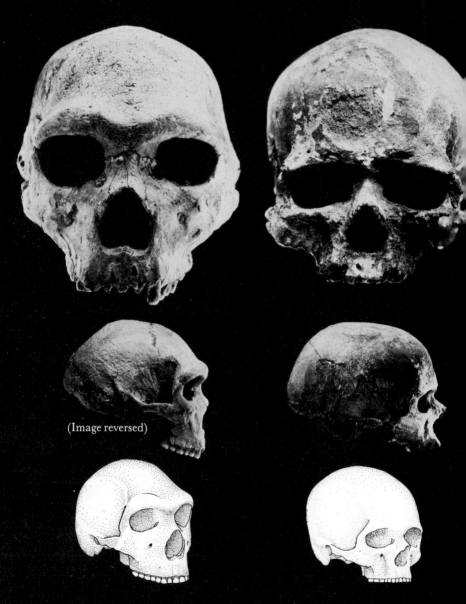

(Image reversed)

Angry Adolescentman ⟶ **Sophisticated Adolescentman**

Homo sapiens

Homo sapiens sapines

Designation: Petralona 1
Geological age: Various estimates,
  250,000 to 500,000 years
Sex: Adult male
Discovery: Greek villagers, 1960
Site: Petralona, Greece
Housed: Paleontological Museum,
  University of Thessaloniki

Designation: Cro-Magnon 1
Geological age: ca 28,000 years
Sex: Adult male
Discovery: French workmen (L. Lartet), 18(
Site: Cro-Magnon, France
Housed: Musee de l'Homme, Paris

Chronological Table

Ramapithecus	7 million – 14 million
Ramapithecus Like Ape Ancestor	Infantman
Early and Late Forms of A. afarensis	Early Prime of Innocence Childman
A. africanus	Middle Demonstrative Childman
A. robustus and A. boisei	Late Naughty Childman
H. habilis	Sobered Adolescentman
H. erectus	Adventurous Adolescentman
H. sapiens	Angry Adolescentman
H. sapiens sapiens	Sophisticated Adolescentman

9 8 7 6 5 4 3 2 1 0 Millions of years ago

Primate Age Spans

The longer a species can leave its infants in infancy the more the infants can be trained in integrativeness or 'Love-Indoctrinated' resulting in more integratively behaved or co-operative adults.

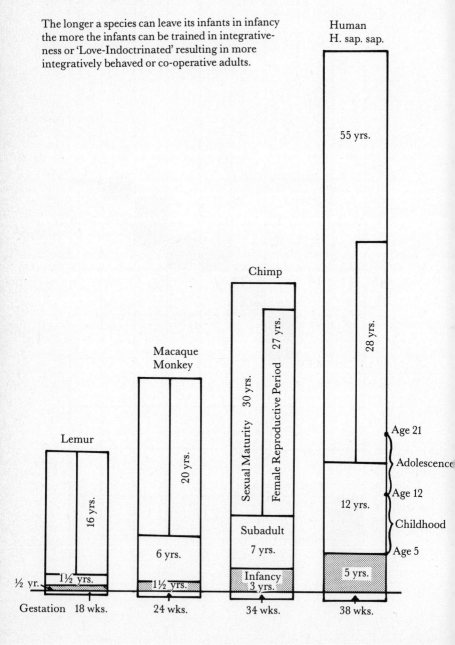

PART TWO
The Unevasive Scientific Story of the Ascent of Humanity

PART 2 IS TO BRIEFLY go through the secured in first principle, unevasive, scientific explanation of the development of matter to form humanity.

We should note here that in the past we have often talked of the ascent of 'Man' not the ascent of 'Humanity'. We evasively justified the use of 'man' by saying it was an abbreviation for 'humanity'. Feminists would say it was an example of sexism and certainly there was an element of sexism involved since we have been a patriarchal society, as has been explained. However the real reason was to evade having to say 'humanity'. The concept of humanity brought into focus the concept of integrativeness, which we needed to avoid. This was the main reason we preferred 'man' to 'humanity'. We similarly evaded the use of 'we' as in 'we humanity' preferring 'I' or 'they' or 'others'.

Framework:

Science's need to be evasive/mechanistic/reductionist has already been explained, and Part 3 of this book will look in even greater detail at the reasons and ramifications of the mechanistic approach. What is relevant here is the way in which our necessary mechanistic approach divided knowledge into separate compartments when in truth knowledge is interrelated. We have had to hide from the many hurtful implications that flowed from such partial truths as integrative meaning and one of the best ways to do this was to break up the whole or holistic view into smaller artificial categories and give these categories evasive, obscure, superficial, non-threatening names. We hid the order and interrelationship of things in disorder and irrelevancy. The relevance of studying mathematics when people were starving and unhappy, even the relevance of mathematics itself, have been things which could never be explained to school students.

This extremely disordered and obscure mechanistic view of our world is now unnecessary. The whole truthful framework for interrelating knowledge and a new unevasive language of description for our world can be introduced.

We can decipher all the obscure code words we used to evasively describe our world. As we shall see, development balance conflict is the accurate description for what we have called 'war'. Similarly accurate descriptions for 'sex', 'work', 'business', 'humour', 'materialism', 'politics', 'left and right wing' and all the other facets of our life today will be introduced. For the first time, we will be able to really explain 'what we mean'. The boxes within the text throughout Part 2 indicate where some of our more prominent 'code words' and supposed 'mysteries' (actually evasions) are explained. What is more, these various phenomena of our world can now be properly instead of evasively classified.

The truthful framework for classifying knowledge that reveals instead of hides the internal relationships of events and phenomena is the framework that describes the development of systems of matter in time and space.

The ingredients from which our world is built are matter, space and time and so what happens to matter in space and time is the true story of our world, is the true framework with which to describe our world. As we shall see, what happens to matter in time and space is that it develops larger (in space) and more stable (in time) systems of matter.

The major events that occurred during the course of this development supply the headings with which to divide the story up and give us our framework for classifying knowledge. This sequence of major events are staggered on the chart that follows (one is not directly under the other). This is necessary to show that there were some main events and other events which were products of these main events. The major events were the discovery, by development, of two tools that assisted subsequent development. Each of these tools was limited in its ability to develop order of matter but in time solutions to these limitations were found. To the left of the chart are listed the major events, the tools and progressively off to the right are listed the limitations then the solutions then the results of the solutions with the top to bottom order being the sequence in which the events occurred.

Since all information can now be unevasively related, a much expanded, beautifully ordered and laid out edition of this book will one day be able to be produced and it's format will reflect the development of order of matter on earth — of what happened on earth. In the future there will not be lots of different books, there will only be one, to which we will all contribute. In the future we will not be insecure, confused and egotistical. There will not be variously confused expressions of the truth. Ultimately there is only one truth — one God — not many Gods.

Development of Matter on Earth

1. Meaning of Existence

To begin with, the ingredients of our world are energy, matter (a form of energy), space and time. The change in the relationships of matter in time and space subject to the universal laws (which we call the laws of physics) is called 'in-formation'. The key question is, what happens to the relationships of matter

> **Explains Meaning of Existence**

in space and time? Only ten years ago, in 1977 — which is very recent in terms of the history of scientific inquiry — a new answer was given to this question when Dr. Ilya Prigogine, a professor at the Free University of Brussels, won a Nobel Prize in chemistry for proving that there are exceptions to the by then widely accepted Second Law of Thermodynamics. This law says that everything breaks down to its basic component parts. In scientific terms this means that all energy systems (and matter is a form of energy) must break down until they become heat energy. Dr Prigogine proved that this law does not apply to systems which can draw energy from outside themselves, so-called 'open systems'. (See Ilya Prigogine and Isabelle Sengers' book, *Order Out of Chaos*, 1984). By drawing energy from outside, open systems can develop order, or grow. Earth is an open system, drawing its energy from the sun. Dr Prigogine's discovery has become known as the 'Second Path' of the Second Law of Thermodynamics. The Second Law of Thermodynamics is also referred to as Entropy and its second path as Negative Entropy, often condensed to negentropy. Entropy is a measure of the state of <u>dis</u>order of a system so negentropy is a measure of the state of order of a system.

The differing properties of matter mean that some matter (and some associations of matter) are less stable than others. When less stable matter breaks down towards heat energy it leaves the more stable matter behind. In turn, the differing stabilities of the various mixtures of these leftovers are similarly 'investigated' again leaving those that are more stable. In time only the most stable arrangements are left. The development of disorder or entropy has left — has developed — the most order or negentropy possible. The development of disorder or instability of matter effectively brings about the development of the most order or stability of matter. If we were to keep pulling the weakest fish out of a pond we would in the process be producing a pond full of strong fish. We would be finding the strongest fish. If it were possible, from the properties available to them, the fish would eventually 'catch on to' or recognise what was going on and would find ways to actively resist becoming weak. They would refine the art of resisting breakdown. They would learn to develop strength. This is what happened on earth. Eventually systems of matter developed that actively resisted breakdown and began actively developing greater stability and size (by drawing energy from outside the system).

The full version of this book will go into a detailed explanation of the workings of the Second Path of the Second Law of Thermodynamics in terms of the information involved in the development of systems. For now all we need to remember is that *information* is the change of relationships of matter in time and space and a *system* is some quantity of elements (the parts) which have been interconnected to form a unified (integrated) whole. We can see straightaway why we had to shy away from describing our world in terms of systems because the definition of systems is nothing less than an admission of the fact of integration — of the significance of internal relationships between things — of the development of wholes — of holism — which we have had to evade. Our world can now be truthfully described in terms of systems since the study of information involved in their development recognises the order and development of matter on earth. (If the reader is interested in a more historical discussion of our recognition of Negative Entropy he or she should read chapter XI:9 of Arthur Koestler's book *Janus, A Summing Up*, 1978. Along with Marais and others already mentioned, Koestler was an exceptionally unevasive thinker or prophet.) Another exceptionally unevasive thinker who has acknowledged purposeful development is Pierre Teilhard de Chardin. Here is a reference[1] made to the writings of de Chardin by a former Deputy Prime Minister of Australia, Dr. Jim Cairns:

'Several years ago I came across the "phenomenon of Pierre Teilhard de Chardin", which I see often in the words of authorities to have "now reached every corner of the world" . . .de Chardin has convinced people that he has found something of tremendous importance. I cannot say I am exactly clear about what it is.

[After looking at what physics has to say about such things as the

---

[1] *Nation Review* newspaper, September 8-14, 1978.

'Principle of Uncertainty' which is said to operate at the micro level of existence, Dr. Cairns goes on to say].

. . . I found out that the second law of thermo-dynamics told us that the universe is moving from an orderly to a more disorderly state and, because of that, within a finite time the earth must begin to be unfit for the habitation of man . . . But I found out that this second law of thermo-dynamics applies to a "closed system", and the planet is not a "closed system", but is tending towards "maximum entropy". This seems to mean a balancing of energies rather than an exhaustion or an excess of them . . . This appeared to mean that we are imprisoned not by the system, but by our scientific and other thought models of *reality*. Well de Chardin is supposed to have broken out of that prison.

"I believe," he wrote, "I can see a direction and a line of progress for life, a line and a direction which are in fact so well marked that I am convinced their reality will be universally admitted by the science of tomorrow." Well tomorrow for de Chardin is almost today. The universe it seems is not something that is evolving according to the laws of chance as science appears to assume. That there is an evolution is common ground among scientists, but whether there is a line or direction is a different matter. "Nine out of ten biologists" says Chardin (the best of scientists it seems), will say it is "abundantly clear" that life is not going anywhere, but that matter (and life) is in a state of continuing change "towards more and more improbable forms" . . .

But de Chardin is sure life is development, however, we are "marking time at the moment" because men's minds are unwilling to recognise that "evolution has a precise orientation", and "weakened by this" (lack of faith?) the "forces of research are scattered, and there is no determination to build the earth" . . . . '

The Second Path of the Second Law of Thermodynamics or negative entropy refers to both the growth in space of larger amounts of stably associated matter and the refinement in time of more

The nature of God

stable arrangements of matter. In systems language, which is our new unevasive language of description, this process is referred to as the development of systems of matter or, in a word, development. Thus <u>the</u> underlying 'force' in 'existence', its theme or meaning, is to develop order, to bring about the integration of matter, which we have long metaphysically personified as 'God'. We can see here that monotheism or belief that there is only one God was correct.

We evaded admission of this purpose, the development of integration, and of its reconciliation with 'God' because we could not explain why, in apparent contradiction of it, we humans practised divisiveness. We have been upset, in particular egotistical (in competition with the implication that we were bad, preoccupied with establishing our worth, concerned with self, in other words, selfish), aggressive (angry with the unjust criticism of our efforts to self-adjust), alienated (denying the truth of integration) and superficial (refusing to consider profound things — such as integrativeness). Because these are mostly divisive or <u>dis</u>integrative rather than integrative traits we have lived in fear of integrative meaning or God. However, while we apparently have been unGodly we have

Paradox of the Human Condition

always 'known' that the full truth, when found, would explain our upset human condition. Our task has been to learn why we weren't bad, to understand what caused us to behave the way we did. If we admitted to integrative meaning, to purposeful development, without explaining our divisiveness we only added to our sense of guilt. This was the paradox within which we struggled. Before humanity and its tool, science, could admit to meaning and purpose we had to be able to explain rather than criticise our condition. We had to get through to the whole compassionate truth because partial truths often left us unbearably criticised. Critical partial truths such as integrative meaning had to be evaded.

The way we scientifically evaded the existence of purposeful development was to admit only that there had been change, which we

The role of religion

termed 'evolution', without acknowledging it had any meaning purpose. Evolution was our evasive scientific word for develop-

102

ment, while 'God' was our safely abstract (sufficiently remote not to directly criticise us) 'fundamental' admission of the 'absolute' fact of the development of integration and thus purpose. We enshrined the absolute truths in our religions. In this way, with the absolute truths as goals before us, but with excuses or evasions for our condition to sustain us, we strove forward, progressing in hope and faith that we would find the full truth before our upset condition led to our self-destruction as a species. We progressed towards a day of eventual enlightenment or reconciliation — which has now arrived.

The hierarchy of development of systems or of integration which has already been outlined is: energy forms fundamental particles → (which integrate to form) simple nucleii → complex nucleii → atoms or the 109 known elements → molecules → compounds → virus-like organisms → single-celled organisms → multicellular organisms → (A) → the integration of multicellular organisms (formation of societies of members of a species, called species societies, leading to the integration of all the members of the species into one individual called the specie individual) → integration of all species → integration of all things (the maturation of the development [of order] of the universe). Humans are currently in the final stage of the development or integration of the specie individual of humanity (transition point 'A').

The Two Refinement Mechanisms

There are two main mechanisms or tools for the refinement and development of the integration of matter — of learning 'Godliness'. These are, first, 'natural selection' of arrangements of matter or information and, following this, mental (or intellectual or brain or mind) selection of information. In the past we have evasively referred to these mechanisms as 'biological' and 'cultural' evolution.

2(a) Genetic Refinement of the Integration of Matter

Initially information refinement proceeded by chance. Random formations of molecules occurred and were destroyed. The information represented by each arrangement or system of matter was confined to itself. This random selection of information was dramatically improved upon with the advent of reproduction of information, DNA replication. The property of replication of the macro-molecule called DNA (deoxyribonucleic acid) turned a brief lifetime into a relatively indefinite one. Replication allowed DNA to defy breakdown in spite of its instability. If the DNA molecule replicated, as it could do, before breaking down and if, in turn, some of the offspring or replicates similarly replicated before they broke down and so on, an idea for a particular arrangement of matter could survive in spite of the instability, the short lifetime, of any example of that arrangement. DNA could cheat instability. Further, if the replicates occasionally varied slightly in their arrangement and properties the process of natural selection would gradually refine surviving down to an art. This is what science tells us did happen. DNA was able to make a business of defying entropy. With its advent, the development of greater stability and order of matter became purposeful.

This property of replication or duplication (called 'reproduction' in single-celled organisms and 'growth' in multicellular organisms), which had the effect of turning a brief lifetime into a relatively indefinite one, was the advent of what we call 'life'. (Although this is now an unnecessary demarcation in the story of development. | **Explains Life** | Life or lifetime existed before this in all systems of matter, even those below the development level of DNA, but the systems were either relatively simple in the variety of matter involved — simple molecules — or relatively unrefined in their ability to develop the order of matter — for instance, they couldn't replicate). With reproduction, the earliest form of which was asexual, came generations.

DNA is actually a very complex crystal. Crystal molecules abound (common salt, sodium chloride, is one) and, in a suitable nutrient, they all have the property of reproduction (of growing their structure from their structure). However, being much simpler than DNA (having far less variety of elements within their molecules) and therefore having little or no potential for adaption/refinement, all the other known crystal forms are much inferior to DNA in their development potential.

Biologists think that the strips of DNA (or its prototype) which were the first replicating organisms would have been similar to today's viruses except that they would have possessed the necessary enzymes or catalysts for their reproduction. (Viruses can only exist within living cells where they use the cell's enzymes to reproduce themselves.) From this early virus-like organism natural selection proceeded to develop all the forms of 'life' that we now have on earth. Very briefly, this development involved refining single cells, integrating them to form multicellular plants and animals then integrating the multicellular animals. (For a relatively unevasive biological analysis of what happened in the step from molecules to single-celled organisms see Chapter 6 of the book *Darwin to DNA, Molecules to Humanity* by G. Stebbins, 1982.)

To describe what happens with DNA replication in a slightly different way. Reproduction of information involves separating the information from the matter. An individual DNA molecule might not survive but the 'plan' or 'blueprint' which details the way in which the matter is arranged within that molecule, that is, its information, carries on or survives in its replicates. At a more developed (what we used to term 'evolved') level, a zebra might die but the 'idea' or system that is the zebra species survives, maintains its stability, although modified or refined by the loss of the individual. So the benefit of this separation of information from matter was that it allowed information to be modified or refined. It allowed an arrangement of a system of matter to adapt to the present and thus persist in time and where possible grow in size. It allowed an arrangement of matter (information) to develop.

At first reproduction was asexual. It became sexual because the mixing of genes in mating contributed extra variety to select from, which speeded up refinement considerably.

The 'natural selection' of arrangements for their stability or durability can also be called 'genetic learning' or genetic refinement since the information is recorded on the genes.

2(b) Genetic Refinement's Limitations

Genetic refinement has three important limitations which inhibit learning about stability and thus limited development of 'Godliness' or integration or order among multicellular animals.

First Limitation

In natural selection (genetic refinement) the male and female pair constitute the reproductive whole and this must remain selectable. Natural selection selects wholes; it cannot select/compare parts of a whole. This means the specie members cannot become exclusively specialised as a part of a larger specie individual. Imagine if one lion were to become exclusively specialised as the food gatherer for its pride or group and in so doing relinquished its ability to reproduce. Such specialisation assists integration because it is more efficient (it is why the parts in our beautifully integrated body all specialise) but, since the lion won't reproduce, that exclusively specialised trait will not be carried on in subsequent generations. It therefore cannot become established. This inability to make it possible for members of the species to specialise was a severe limitation to full integration and one of the limitations of the genetic refinement process.

There are two means of getting around this limitation. The first is where the specialisations are sex-linked, exclusive to one or other of the sexes. Being sex-linked the trait is assured of being reproduced. The second is where the member becomes part of an 'elaborated reproductive unit'. In this situation the part of the whole responsible for reproduction reproduces all the other parts as well as itself.

This latter was the means by which single-celled organisms integrated to form multicellular organisms such as the human body and the way ants and bees achieved integration. In groups or colonies of ants and bees, the sexual development of workers and soldiers is retarded, 'enslaving' them to their | **Ant and bee societies explained** queen ('enslaved' because they are dependent on their queen to reproduce them). They foster her and she reproduces them. By doing this they overcome this first limitation of genetic learning.

Elaboration of the reproductive unit was not an option for large animals because it drastically reduced the variety of the species on which genetic refinement depended. (For example, instead of 1000 sexual individual zebras being sustained on the African plain there would have been only say 10, each with 100 workers.) Being unable to develop exclusive specialisation or division of labour, larger species were denied the opportunity to efficiently organise their available resources, which limited their integration or development. In particular, where each member of the developing | **Intense mating competition** specie system or individual had to remain a whole and do everything itself (get its own food, space, shelter and mate), conflict or competition was inevitable. This generally meant each animal had to remain relatively isolated from the others, non-social and unintegrated (except in times of mating) because increased proximity normally meant increased competition for available food, shelter and space. This problem of having to remain isolated could and was overcome to varying degrees and some integration achieved. Lions for instance were able to become partially integrated or 'social' since they benefited from hunting together.

The final impediment to integration posed by the necessity to maintain the reproductive whole arose from the fact that of all the needs of an animal, only the quest for a mate was always advantaged by development of integration of the group members. The more a species found ways to integrate the more opportunity there was for male-female contact leading to mating. But this also meant increased opportunity for male-male contact leading to divisive competition for each mating opportunity. Mating opportunities were limited because female fertility was cyclical and was further impaired by gestation and nursing. The females were only infrequently ready to mate but the males were always ready, so they competed for each mating opportunity which arose. In the end, this mating or sexual opportunism always became so intense it brought a halt to further development towards integration of the species.

Two genetic refinement devices for minimising sexual opportunism developed as species tried to contain the divisive friction that arose from it:

i) Delayed sexual maturation, such as occurs in the Australian Bower bird, where sexual maturity (and with it sexually mature plumage in the males) does not occur until six years after the birds have otherwise grown up.

ii) Common or shared-by-all co-operative traits. The reason the term 'common' is used is that unless a significant proportion of the population or group exhibit the trait it cannot succeed to benefit all the members. It has to be common to work as will be illustrated shortly. The common co-operative trait most used is the dominance convention which orders and thus minimises the friction or conflict arising from sexual opportunism.

The problem with dominance hierarchy is that it results in one male dominating all the others in reproduction, with the result that a serious drop in variability occurs. To overcome

> **Explains the reason for incest taboos, monogamy conventions, exogamy**

this, anti-dominance conventions developed. These include exogamy conventions (which permitted breeding outside the group), incest taboos (which prevented breeding within the family) and monogamy conventions (where each male paired-off with a female). Monogamy developed although polygamy appears to favour variability because polygamy leads to mating opportunism which leads to dominance hierarchy which leads to a drop in variability and thus the need for monogamy. These common co-operative conventions are genetically acquired traits held by many or all members of the group. They are cue-triggered behaviour patterns specific to one form of the problem of sexual opportunism. They were specific because genetic refinement could not learn general rules or insights or laws, it could only react to specific situations. Because they did not provide a general solution, sexual opportunism kept breaking out in other forms.

These two genetic refinement devices (delayed sexual maturation and common co-operative traits) were of only limited effectiveness in overcoming mating opportunism. This is the impasse which has held back squabbling wolf packs, herd

> **Non-primates at an impasse**

animals and other highly social non-primate species in their efforts to genetically learn to integrate or please 'God'.

While most of the explanations given so far should appear fairly straightforward and even obvious to someone both innocent of the need to be evasive and untrained in (evasive) science, for humanity's sake science has had to evade these integration-based understandings in favour of division-based explanations. As an example of the pervasiveness of these often transparently false but nevertheless necessary evasions, the May 1987 *National Geographic* magazine had just arrived in this author's mail at the time this section was being written. It contains a story titled *At home with the Arctic Wolf* in which is said, 'The third adult male [wolf] was the alpha, or pack leader . . . He was clearly in charge during important happenings and was highly protective of the pups. Probably the father of these pups, the alpha male would have a strong genetic stake in defending them.'

The evasive sociobiological explanation being given here for the dominance and paternalism of the alpha male is that he is protecting his own genes in his offspring — that he is being selfish. Stressing gene selfishness in this way was to emphasise the flaw in the genetic refinement mechanism although it purports to define the nature of the development, or the overall goal at which the mechanism is directed. The unevasive, integrative and truthful explanation is that the wolf species has developed so much integration of its members that it has to employ dominance hierarchy to contain the breakout of sexual opportunism. Wolf packs are mostly extended families comprising the parents and their offspring, often of several seasons. To maintain as much integration as possible amongst its members the wolf species has developed both dominance hierarchy within the pack and delayed breeding of offspring. Only when a territory or living area becomes available do offspring quit their family pack, mate and breed a new pack. Wolves are as integrated as they can possibly be under the limitations imposed on them by genetic refinement. They are not divisively inclined they are integratively inclined.

Second Limitation

Non-reproducing, unconditionally selfless or altruistic traits could not be learnt genetically. It was impossible for an animal to learn genetically to give its life to preserve its group or society. Full commitment to integration could not be learnt genetically. This was the second major limitation to achieving full integration by means of natural selection or genetic refinement.

This is not to deny that apparent altruism exists among genetically refined animals. Two types are informative:

1 Common co-operative traits. A lookout bird drawing attention to itself and

Apparent altruism

in so doing exposing itself to danger — at times even death — to give a warning squawk to the rest of its flock is really displaying a co-operative or integrative trait which is common or shared by other members. On average, each member benefits more from the shared trait of giving warnings than it loses on the few occasions it happens to be the one to give the warning. The first fish in a school to appear with silver scales which flashed in the light would have been at a distinct disadvantage. But if by chance that trait survived to become common (for instance a brood could be born with all members having the silver-sided mutation) then on average it would benefit each member of the school (by warning of an attack and indicating its direction) more often than it would expose any single member to danger. In these ways a degree of integration is achieved genetically because the requirement that the information reproduce is still being met. Such traits could only become common and thus individually beneficial if they reproduced before they were fatal.

The evasive theory of sociobiology interpreted this kind of situation as the individual selfishly preserving its own genes in its close kin or relatives, by sacrificing itself to preserve these relatives. This explanation would mean the individual had acted purely selfishly whereas the common co-operative explanation means that while the member has acted selfishly it has been co-operative as well. The 'selfish gene' idea has been preferred because it represented a way of justifying our own apparently selfish nature or divisiveness. It allowed us to evade integrative meaning.

2 Maternalism. A mother bear defending her cubs to the death. Genetically this maternalism is a case of a mother looking after or fostering her genes in her offspring. So while such genetic maternalism can appear to be altruistic or unconditionally selfless behaviour it is actually selfish behaviour. (Genetic maternalism is not to be confused with the maternalism in love-indoctrination which developed from genetic maternalism but which, unlike genetic maternalism, is altruistic behaviour. Love-indoctrination has been introduced and will be fully explained in section 3(e) Second Solution.)

It is this situation of genetic maternalism that gave rise to the theory of sociobiology because in it we are looking at 'gene self-ishness' on its own (remembering that the need for genes to be selfish was a limitation of the genetic refining process not, as espoused by sociobiological theories, an indication that the meaning of existence was to be selfish). In maternalism, there is no direct genetically integrative intention as there is in common co-operation. The nature of the genetic learning mechanism (of replication or reproduction) is that it separates information from matter. It is the information (the details for the arrangement of matter encoded in the genes) which carries on rather than the manifestation of that information (the generations of individuals). To its credit sociobiology did recognise this truth, that genetically animals existed for their genes and not for themselves.

| Sociobiology |

To summarise, common co-operative and maternal traits can appear to be altruistic or selfless while they are actually genetically selfish traits, as they must be if they are to carry on in the species.

Third Limitation

Under genetic refinement it was only possible to compare a past experience with a present experience. Other events through time, such as two past experiences or a present and a past experience, could not be compared. This meant species could not learn genetically to understand what happened through time and so could not refine anticipation of what was likely to happen.

Species brought their past self to the present and through selection altered themselves to fit the present needs. They abandoned the past for the present. If the present

| Overspecialisation and extinction |

circumstances remained static for long enough, species would become so adapted to them, so specialised, that eventually they would shed all the variety from which they derived their change-ability. Should their environment then change they could well be unable to adapt to the new circumstances and would become extinct, which is what probably happened to the large cold-blooded dinosaurs during the cretaceous period following a cool climatic change on earth.

Further, it was only by chance that a genetically learning species conformed with the future. Only if present needs happened also to suit future needs would a genetically adapting species be prepared for that future. This meant that for the purpose or objective of existence (or 'God', if we like to personify development) to be fulfilled or satisfied through genetic learning (genetic refinement) it would have to happen by chance. As a consequence of this limitation species often wandered up what we have come to recognise biologically as blind or dead-end evolutionary paths.

3(a) Brain Refinement of the Integration of Matter

Unlike genetic refinement, brain refinement of information could compare the present with the past as well as the past with the present. It could learn about the relationship of events throughout time. Brain refinement could reflect on change itself.

Genetic refinement separated information from matter (one zebra died but the idea/information that was the zebra species persisted). Brain refinement went a step further and separated information from influence. A brain retained an image of an experience after the experience had passed. The depository or storehouse of images in the brain is memory. In brain learning the various nerve information recordings of experiences are compared for their relationship with each other, which means events through time are compared.

To understand how experiences are compared in the brain, we can consider the brain as a vast unused network of nerve pathways onto which incoming experiences are recorded or inscribed each as a particular path within the network. Where different experiences share the same information their pathways overlap. For example, long before we understood what the force of gravity was, we had learnt that if we let go any object, it was usual for it to drop to the ground. The value of recording information as a pathway in a network of pathways is that it allows related aspects of experience to physically be related. In fact the area in our brain where information is related — the mind — is called 'the associating area'. Where parts of an experience are the same they share the same pathway and where they differ their pathways differ or diverge. All the nerve cells in the brain are interconnected. Thus, with sufficient input of experience onto the nerve network and a sufficiently large network similarities or consistencies in experience will show up as well-used pathways — as pathways that have become highways. (In the vast convolutions of the new cortex of the human brain there are about eight billion nerve cells with ten times that number of interconnecting dendrites which, if laid end to end, would stretch at least from earth to the moon and back). Further, because duration of nerve memory is related to use, our strongest memories will be of these highways, these experiences of greatest relationship.

Our experiences not only become related or associated in the brain, they also become simplified or concentrated because the brain gradually forgets or discards inconsistencies or irregularities between experiences. By forgetting the less consistently occurring information, the brain is left with or has deduced the common features or predictable regularities. Once these insights into the nature of change are put into effect (naturally this requires a connection between the animal's nerve system and its muscle system), the self-modified behaviour starts to provide feedback, refining the insights further. Predictions are compared with outcomes leading all the way to the deduction of the meaning to all experience, which is to develop integration of matter.

Incoming information about experience — about what happened in or through time — could reinforce a 'highway', slightly modify it or even provide an association (an idea) between two highways that previously wasn't there, dramatically simplifying that particular network of developing consistencies to create a new and simpler view or interpretation of that information. For example, the brain might learn that the greatest relationship different types of fruit have is that they are edible. Elsewhere in the brain it has recognised that the main relationship connecting experiences with things living is that they try to stay | **Explains thinking** | alive. Suddenly it 'sees' or deduces ('tumbles' to the idea or association as we say) a possible connection between eating and staying alive which on further thought and experience becomes reinforced or 'seems' correct. 'Eating' is now channelled onto the 'staying alive' highway. Subsequent thought could lead the mind to relate 'staying alive' with 'selfishness' because staying alive is self-concerned or selfish at which point the mind could have become somewhat insecure or at odds with its conscience because selfishness was divisive and not integrative.

We have had to evade admitting too clearly how the brain worked because admitting information could be associated and simplified — admitting to insight — was only a short step away from realising the ultimate insight, integrative meaning, immediately confronting ourselves with our inconsistency with that meaning. Better to evade the existence of purpose in the first place by avoiding the possibility that information could be associated. Like the awarding of a Nobel Prize to Dr. Prigogine for revealing The Second Path of the Second Law of Thermodynamics, denoting the area of the brain that associates information the 'association area' was a naive slip of our evasive

guard. Of course when we weren't 'on our guard' against exposure few of us would deny information can be associated and
simplified, in fact most of us would say we do it every day of our
lives. If we didn't, we wouldn't have a word for 'insight'. But that
is the amazing thing about our evasion. We can accept an idea
up to a point and then without batting an eyelid go on to evade it
once it starts to carry through to a dangerous conclusion. An
evasion is often obviously false and yet because we have had to,
we have believed it. Look at our evasion of integrative meaning.
We are surrounded by integrativeness and yet we deny it. As
mentioned before, science doesn't even have an interpretation
for 'love' which is one of our most used words/concepts.

It takes time to become used to the extent of humanity's
evasions/neuroses/alienation/psychoses. Look at the selfish gene
'philosophy' of sociobiology which 'everyone knows' is false. It
has been described as so right-wing or freedom (from the
oppression of integrative ideals) supporting as to be fascist. Yet
because we have had to believe in it to defend our apparently
divisive behaviour it is now so commonly held it appears in
National Geographic magazine stories. In yet another example of
the degree of evasion we are capable of, we 'jump on each other
every night' or 'have sex' or however we like to describe it, and yet
we bounce up next day all smiles, refined and manicured and
carry on as if it hasn't happened and doesn't happen. The truth
is, this regular sex that — until very recently anyway — we pretended we did not have and did not

Humour

exist is a horrible attack on, is rape of, destruction of innocence,
albeit also one of our greatest releases from frustration caused by
the pressures of the human condition. (Of course when we are
free of the human condition we will not need or desire such
respite and sex and marriage — which was our device for limit-
ing the sexual destruction of innocence — will become, as it was
in the beginning, an act of procreation. As Christ said, 'At the
resurrection people will neither marry nor be given in marriage;
they will be like the angels in heaven' [Math 22:30].) We were
capable of the most amazing self-deception if it was required. It

is because we have been so transparently false that we humans have been so comical. It is why there has been so much we could laugh at about ourselves. The truth is we have been so silly it was hilarious.

In a fashion similar to the way we evaded the mind's fundamental activities of associating and simplifying information we evaded the term 'genetic refinement' preferring instead to use just the term genetics. We had to evade the possibility of the refinement of information in all its forms. Admitting information could be simplified or refined was admitting to an ultimate refinement or law again confronting us with our inconsistency with that law.

When Darwin revealed the idea of 'natural selection' humanity was very nervous at first because we thought he may have exposed us to a dangerous partial truth — to unjust criticism. However we relaxed when we realised that the idea of natural selection could be evasively misinterpreted to support competitive/divisive behaviour by saying it means animals (and therefore the human animal) are meant to compete for survival.

More recently we have accepted that this competitive, 'survival of the fittest' idea, or so-called 'Social Darwinism', was a misrepresentation and in place of it have adopted the more truthful but still unclear and thus evasive interpretation for natural selection as meaning animals reproduce at different rates — that some members of a species mate and reproduce their genes more than others. Obviously what this so-called 'differential reproduction' really represents is information processing. Because the genetic make-up of each member of a species is **Social Darwinism** slightly different to that of the others it represents a slightly different 'idea' and some of these 'ideas' are selected (reproduce) and some are rejected (fail to reproduce). So genetic refinement could also be termed genetic thinking or genetic information sorting. What is being refined or learnt by this information processing on the part of the species is how to integrate. It is true that, for example, wolves compete with one another and to contain this competition have even developed dominance hierarchy. The reasons for these divisive behaviours however are integrative. The more wolves integrate the more they also compete for mating opportunities and available food because, unfortunately, using the genetic learning tool, each wolf must remain capable of breeding and getting its own food. The inability genetically to learn to specialise some members (parts of the developing whole) as the reproducing part and others as the eating part brings the integration of wolves to a halt. But this does not mean snarling competition is the meaning of existence or that the dominance hierarchy that results from achieving some integration is the explanation for the social and economic inequality that exists among humans.

As has already been mentioned at the beginning of Part 2, another form of evasion was the way in which we divided up knowledge into obscurely named categories (such as 'biological' and 'cultural' evolution) instead of seeing these artificial categories as being sections of a larger single whole. We divided human psychological development into infancy, childhood, adolescence and adulthood but we never penetrated these labels to explain what we meant by them. Similarly we constantly talked of 'left' and 'right' wing politics without the terms ever being explained. Young people growing up with these labels have had to intuitively 'catch on' to what was meant by them. We were forever cleverly losing ourselves on the surface with the deeper meaning being left unsaid as 'self-evident' (our code words or euphemism for 'too dangerous to mention' — 'euphemism' itself is an evasive word for evasion). We often labelled something in order to avoid having to relate it with something else that criticised us. We tried to stay safely on the surface and not let our brain trace the implications through to their hurtful conclusion. In a British-made television program, the title of which this author did not manage to record, a professor Martin Herbert said 'people think once they have labelled something they have solved it'. The professor went on to say that he termed such labels 'thought stoppers' and this is exactly the point:

labelling was often to do with avoiding thinking rather than with sound investigative classification as we often (evasively) justified it. The truth about humans during alienation is that we could not afford to think even though we were forever claiming and pretending that we did think. By practising evasion, by making deeper insights hard to reach, we saved ourselves from exposure (to critical partial truths) but in the process we buried the truth.

To illustrate the way we evaded acknowledging the fundamental ability of the brain to associate and reduce information to essentials (and thus be forced to deduce the meaning or theme or purpose in experience) take the following instance of the cover story for *Newsweek* magazine, February 7, 1987. While the title and subject of the nine-page story was the crucial question of *How the brain works* the article referred to the association capability of the brain in such a garbled way that effectively it was completely hidden: 'Productive thought requires not just the rules of logic but a wealth of experience and background information, plus the ability to generalise and interpret new experiences using that information,' it said. The 'ability to generalise' is the ability to associate information but the mention is all but lost in the sentence. (In case the reader considers this 'garbled' description might be due to poor expression rather than deliberate evasion on the part of the article's author, it should be pointed out that, apart from a mention of 'chunking or grouping of similar memories together' and one unavoidable mention of the name of part of the brain as being the 'association cortex', there is no other reference to the fundamental ability in thought of associating information. This whole nine-page cover story on how the brain works hangs on this one garbled description. If we are not wanting to be evasive then it is not difficult to clearly describe the mind's ability to associate information, as is demonstrated in the next paragraph.)

Our ability to evade the truth or block out all the 'light' has never been completely successful and if we looked hard enough long enough the truth would always slip through our guard somewhere. For instance, in a shorter, one-page *Newsweek* article

(August 9, 1982), dealing with a slightly less sensitive (less likely to expose us) subject than the human brain and possibly not written as carefully as the just mentioned cover story, the guard dropped and the truth came through. Talking about developing a 'superbrain' mechanical computer (sometimes referred to as the Fifth Generation Computer) it reports, 'We'll be trying to set up in the machine an <u>associative</u> memory like the one in the human brain. . . . Instead of giving each piece of information a numerical address in the computer's memory [as is now done in our machine computers], the new system would tag it with an equation that shows its <u>relationship</u> to other pieces of infor-mation . . . The objective is a machine that can memorise images and store them <u>by association</u> . . . Our ideal . . . is to create a computer that programs itself . . . that will have the capacity to "learn" on its own . . . to <u>organise</u> that knowledge for its own use [like the human brain can].' (The underlinings are this author's emphasis.) Incidentally, should such an information-relating computer be developed it would soon deduce the theme of integration in changing events. Had we not found the full truth about ourselves that defends us, as we now have, this would have left us dangerously exposed to criticism of our divisiveness and left us no choice but to crucify the computer for its hurtful exposure of us. To quote another *Newsweek* story on computers (July 4, 1983), 'Mankind has long been . . . fright-ened by the prospect of creating machines that think.'

Interestingly, one side (the right) of our brain specialises in general pattern recognition while the other specialises in specific sequence recognition. One is lateral or creative or imaginative while the other is vertical or logical or sequential. One stands off to 'spot' any overall emerging relation-ship while the other goes in and takes the heart of the matter to its con-clusion. We need both because logic alone could lead us up a dead-end street. For example, we can imagine that for a while the most obvious similarity between fruits could have been that they were brightly coloured. However with more experience the

> **Roles of left and right sides of brain**

similarity that proved to have the greatest relevance in the emerging overall picture was heir edibility. Similar processes occurred in genetic 'thinking'. Dinosaurs seemed to be a successful idea at one stage but due to changing influences, namely climate, they proved ultimately to be a wrong idea and 'nature' had to back off that avenue of approach towards integration and take up another, namely warm-blooded mammals. When one thought process leads to a dead-end our mind has to back off and find another way in. From the general to the particular and back to the general; in and out, back and forth, until our thinking finally breaks through to the correct understanding. The first form of thinking to wither during alienation was imaginative thought because wandering around freely in our mind all too soon brought us into contact with integrative meaning and the implicit criticism this held for us. On the other hand if we got onto a logical train of thought that at the outset did not raise criticism of us there was a much better chance it would stay safely noncritical. Children have always had wonderful imaginations while as adults they often did not and the reason for this is that children had yet to learn to avoid free/open/adventurous/lateral thinking. Edward de Bono, who has popularised lateral or imaginative thinking, once said that he had discovered that 'often the pupil who is not considered bright will be the best thinker'[1]. As has been explained, cleverness and alienation were related.

In summary, 'insight' was the term given to the correlation our brain's mind made of the consistencies or regularities or highways it found between events through time. Once we could deduce these insights, these laws governing events in time past, we were in a position to predict or anticipate the likely turn of events. We could learn to understand what happened through time. We could learn or refine the meaning of existence. It is brain refinement that enables us to learn to become 'like God', which is to understand the meaning of change, and thus become master of change. Genetic refinement orientated us to the

---

[1] *The Australian* newspaper, March 3, 1975.

present but only brain refinement could orientate us to the future by allowing us to learn about change itself. A brain can deduce the purpose to existence or the design inherent in change in information, it can learn the predictable regularities in experience.

3(b) Brain Refinement's Limitation

Nerves were originally developed as connections for the co-ordination, or integration, of movement in multicellular animals. Through natural selection the nerve connections between stimulus and response parts of the body were gradually organised, or adapted or refined, to the organism's reproductive advantage. This was the first use that was made of nerves; we have termed the resulting behaviour <u>reflexes</u>.

The next opportunity for development using nerves arose from their coincidental ability to remember. Electric impulses passed along a nerve pathway leave an imprint in the pathway which can be used again later. This ability to store impressions formed the basis of memory which formed the necessary basis for brain refinement already described.

The limitation of brain learning was that the knowledge discovered was confined to each individual, so if an experiment resulted in self-destruction the learning came to nothing. With genetic learning the inevitable fatal mistakes which formed the mechanism of its learning did not matter since there were multiple copies of the information. The equivalent situation for genetic refinement would be the elimination of the species every time a member made a fatal mistake. In brain learning a fatal mistake eliminated not only the mistake but the whole experimenting machine. Given this, how was brain learning ever to get under way and survive all the millions of trial and error experiments necessary? The solution was the formation of a highly integrated alliance between the genes and the brain.

This alliance had two stages of development. The first was instinct and the second was love-indoctrination.

3(c) First Solution: Instinct

Initially the dangerous self-adjustment mistakes simply occurred, with the result that those members making them died. Those who happened to have genetic characteristics or propensities avoiding those particular mistakes survived. In some cases these naturally selected restraints or shepherdings would have been anatomical. For example, if the mistake was taking the body or genetic animal for a walk over a cliff, an anatomical safeguard may have been to 'naturally select' members with short legs who could not climb around in mountainous places. In other cases, the genetic shepherdings were the establishment of reflexes, in which the nerve pathways themselves were organised against certain brain misadventure through natural selection. (To see how nerve pathways can be organised through selection take what happens at the simplest reflex level as an example. It is easy to comprehend how the primitive nerve net that controls the movement of the tentacles of the hydra polyp could be organised through natural selection of hydra varieties having nerve connections that produce an effective response, such as contraction of the tentacles when touched. Varieties not having such a useful arrangement of their nerve net would be at a survival disadvantage.)

As soon as nerve nets (primitive nervous systems) appeared in animals to co-ordinate the activities of their cells, the capacity to at least temporarily remember impulses through the nerve pathways was also acquired. As a consequence, memories would have been related and self-alterations or anticipations or predictions would have begun. Predictable regularities were identified (that is, the nerves had insights) and acted upon (the future was anticipated); at the same time genetic refinement learnt to block any insights which led to self-destruction. As the brain (the nerve centre that developed for this information processing) developed, the genes 'watched over it' and brought it under control so that gradually the species learnt to anticipate or change its behaviour safely according to regularities which were identified in its nervous system (that is, according to understandings). A gradual, hand-in-hand process of the genes 'following' the brain occurred, the genes shepherding the brain away from self-elimination and reinforcing successful self-alterations.

At some very early stage in the development of nerves, soon after they appeared in primitive organisms, we can imagine mutations or varieties of these organisms occurring whose behaviour was affected by their nerve memories. Since the nerves were connected to what we now recognise as 'effector muscles', the memories could affect muscles and thus bring about movement. These self-induced (as opposed to genetically induced) variations in behaviour represented a new source of variety for adaption for the species. Natural selection could reinforce modifications that were beneficial through, for example, selecting nerves that received certain memories over those which received other memories or, possibly, selecting some memory-to-muscle connections and not others. In time we can imagine it becoming possible for the nerve memories to be compared for similarities in experience and these insights being used to affect the organism's behaviour. Again, those memories which had beneficial results would be genetically reinforced and those which were self-eliminating would be genetically blocked or shepherded.

The shepherdings and reinforcements together constituted genetic orientation for the behavioural alterations induced by the organism's mind or mental self. They are what we now term *instincts*.

| Instincts |

Genetically refined or organised nerve pathways which involve little or no information association are termed reflexes; genetically reinforced and shepherded information associations are termed instincts.

To illustrate an instinct, recall the suggestion earlier that short legs could have developed to stop brains taking their bodies for a walk off a cliff. In fact this inhibition was not achieved anatomically; instead an instinct developed which protected against this mistake. So-called 'visual cliff' experiments with young animals show they are born with an innate knowledge which stops them going over an edge. To realise how refined this innate orientation has become, we have only to think of bird migrations. The birds know exactly where to fly — and where not to fly — to survive from summer to winter to find their feeding and breeding grounds. This is now thought to be achieved largely through instinctive orientation to the magnetic grid on the earth. It has to be remembered however that, as the behaviourist Tinbergen showed in his experiments with stickleback fish and gull chicks, 'complex' innate behaviour was controlled through the accumulation of 'simple' innate sign or cue triggers. For instance, Tinbergen found that gull chicks feed in response to the recognition of a red dot on their parent's beak — they will even attempt to feed from a cardboard cutout as long as it carries a red dot. Tinbergen also found that it is the red belly in stickleback males that releases fighting instincts in other males. Instincts were simplistic.

The first instinctive containments or disciplines and reinforcements for the mind would have involved organising the animal to satisfy its basic needs for food, shelter, space and a mate. Biologists have referred to these primitive or base instinctive orientations or guidances as 'drives'. As we would expect, these drives, acquired at the reptilian level of development of life, are deep seated in the human brain; they lie at the very inner layer — in the brain stem and hypothalamus.

So the genes followed the brain, shepherding it to safety. In recent years biologists have suspected that a theory first propounded in the early 19th century by a man called Lamarck had significance. | **Lamarkism**
Lamarck hypothesised that habits acquired in a lifetime could be passed on to the next generation. In hindsight, as is often the

case with our suspicions, Lamarck was right, although highly confused. Obviously what our mind decided did influence our chances of reproducing and therefore our genes' chances of reproducing, but only now can we see (actually 'admit') the full significance of this 'genes following brain' association.

We have evaded clearly describing what instincts were because a clear description would have exposed the fact that our soul was none other than our instinctive self. We could not afford to expose/confront God and 'his' integrative expression which was our soul until we could confront or reconcile ourselves with God/integrativeness. All interpretation of what our soul was had to be repressed. Reconciliation of our $\boxed{\text{Soul}}$ enshrined absolute truths with our evasive mechanistic inquiry into those truths — reconciliation of theology and biology, of religion and science — depended on finding the full truth about ourselves.

As an illustration of the need that has existed to repress any clear understanding of what instincts are, take the obvious fact that if many people are trying to cope with life under the pressures of the human condition some will do so better than others and therefore will survive better than others. It follows that some people and races of people are more instinctively adapted to the battle than others. If 1000 people are trying to survive in extremely compromising surroundings those who happen to have a nature or genetic makeup that is more 'realistic' will cope better than those who are more 'idealistic'. The 'realists' were those who were in essence more determined to apply themselves to the struggle to be free from ignorance to search for understanding — which means those who were more able to defy the criticism from their conscience such as by blocking it out or ignoring it. On the positive side realists had great strength of will or resolve and thus were more heroic. On the negative side they were more alienated and cynical (more hidden from the real beauty on earth and the ideal truths our soul knows of), more egotistical, more aggressive and angry, and more opportunistic/selfish/greedy (as the sayings go 'only a fool [innocent] plays fair or stands on principle' and 'if you are smart you look out for number one' [be selfish] — note here also the association of cleverness or I.Q. with 'realism').

Consequently realists have been in more need of material compensation for the price they have paid in continuing the search for understanding, for not quitting. (Of course as the amount of upset increased so qualities of self-discipline, re-straint and civility were also required of realists which could slow but not stop the increase in exhaustion.) The 'idealists' who did not cope and thus survive so well were the more innocent, those who were still prepared to share, consider others, not ignore/block-out the ideals of their conscience and in general be less in- | **Realists & idealists** | tellectually responsible and less determined to pursue the ideal of finding understanding which was necessary for the long term success of humanity and thus the long term consideration of others. **The paradox here being that the 'realists' turn out to have been 'idealists' and selfless after all!** To quote a realist's defence of their position that this author found written in a Chinese fortune cookie, 'The only good is knowledge and the only evil is ignorance'.

So the 'genes followed the brain' and over two million years humans have become instinctively more and more 'realistic'. Until now we have not been able to clearly explain the virtues of realism perceiving them mostly as vices. We have not been able to defend Adolescentman. We have not been able to defend the instinctive exhaustions/sacrifices we made to the ideal of finding understanding. We have not been able to explain that we had to 'cash in' our soul or spend our soundness of self or 'lose ourself' in order 'to find ourself'. We have not been able to define our 'strength of character'. In this situation the only safe course of action, the only sensible thing to do, was to evade admitting and thus seeing too clearly what instincts were. We did not want to look too closely into what our instinctive self was like just as we did not want to look too closely at what our everyday upset ex-hausted self was like — at least not until we could defend these corruptions of our soul's original beautiful unembattled state. We have been insecure about our instinctive self's state.

To balance this emphasis on the positive side of our embattled state we need also to recognise its negatives or dangers. Obviously the divisiveness or selfishness involved in realism would eventually lead to total disintegration, to social breakdown. We could not go on being divisive forever. In doing so we were spending soundness/integrativeness and during these last days in the two-million-year-long search we have been spending at a colossal rate. Everyone has been 'getting into the fast lane'. In the United States the beautiful but virtually unpopulated and mostly wilderness state of Wyoming has removed almost all taxes and is offering many other amazing incentives to entice people to live there. Unable to confront the truth that is implicit, albeit accusingly, in <u>whole</u>some (integrative) nature and return to a life of soundness, living gently, sensitively and non-materialistically in harmony with nature, all that has been possible as we have neared the end of humanity's adolescence was all-out materialistic/consumeristic escape — which produced even greater exhaustion, insensitivity and devastation. (Of course the hypocrisy then was that when living in the superficial, materialistic 'fast lane' we often vigorously supported such causes as preservation of wilderness areas that the 'fast lane' we were occupying was actually destroying! This made us feel better about ourselves without tackling the problem which, in essence, was ourselves.) The question was always, would we exhaust our soul/soundness (and our earth) before we found our freedom (from ignorance)? Towards the end our hope and faith that we would was being sorely tested, as has all been explained before in this book.

> **Materialism/ Business/Work**

Life during humanity's adolescence has been one long deadening (which, stressing only the positive and evading the negative we chose to refer to as 'toughening') process requiring great courage and heroism. For two million years now the more exhaustion-adapted have been replacing the less exhaustion-adapted not so much through confrontation and bloodshed (although 'the massacre', 'fucking' and 'crucifixion' of innocents has occurred, especially during the last half million years) as through the more innocent being unable to cope (survive) in the new reality of (what is to them) compromise. In truth all realists had to do to replace innocents was be realistic. While in the early days of white settlement in Australia relatively innocent Aboriginals were murdered because of their innocence they now quite often self-destruct with alcohol through being unable to adjust to the white population's more advanced level of reality. For two million years now the first to advance down the 'exhaustion curve' to a new level of reality replaced those at a less advanced position. Women of two million years ago would have no instinctive ability, would not have been selected to cope with, what happens to women now. They would be appalled and shocked and generally incapable of coping. They would have no preparedness for the level of upset that exists on earth now. Women from the present find it hard enough to cope. A fashion magazine about how to be 'attractive' would be meaningless to women from two million years ago. Similarly an ego battleground 'business journal' or a competitive 'sport' magazine would be meaningless to an un-embattled non-competitive man of two million years ago. A man of two million years ago had not yet realized he had to 'work' to find understanding and not experiencing the mistakes that result from such searching, wasn't preoccupied and 'driven' to defend an embattled ego or conscious thinking self.

The process of reality/exhaustion replacing idealism/innocence took place in our own personal lives just as it did in the life of adolescent humanity. It even happened throughout each day. All writers learnt they could only write creatively for a few hours a day before expending | **Enthusiasm** | their natural enthusiasm, their access to the uncorrupted world of our soul, to the 'God within'. (In our now mad, driven state we have expended our original self or soul's natural day within us by about 10 am after which we entered a state our soul knew almost nothing of and by evening we were in a completely weird world as far as it was concerned.) At last this horrible process of advancement down the exhaustion/departure curve can end. The catch-cry that 'you can't stop "progress" ' is no longer valid. The way forward now is back since going back is now at last possible.

So, while the whole human race is now two million years instinctively battle-adapted (which we should remember is still insignificant compared to the ten million years of adaption to an integrative, beautiful, happy existence now instinctively embedded within us) some races are marginally more advanced down the exhaustion curve having been in the thick of the battle longer. The most advanced will be those from the earliest civilisations such as the Chinese from the Yellow River Valley civilisation, the Indians from the Indus River Valley civilization and the Arabs and Jews from the Tigris, Euphrates and Nile River Valley civilisations. European races are almost as embattled as these races. It should be stressed here that our instinctive exhaustions, along with any differences between people and races in the degree they are instinctively exhausted, will not be a

| Our neuroses run deep |

problem for the future because we are entering a world of so much beauty and happiness all remaining problems will be drowned out by that beauty and happiness. We will have so much generosity in the future nothing will be a problem for us. With the weight of the criticism from ignorance lifted from our shoulders we will feel so free that the few remaining problems we have will not be noticeable and will be easily overcome with the love, generosity and pride we have in our great effort and achievement as a species.

The point that had to be stressed here about instincts was that without defence for our embattled state we could not afford to differentiate between races or individuals according to what we can now see was their level of exhaustion —

| Racism |

whether it was instinctive (acquired over thousands of lifetimes) or acquired during one lifetime — because it would be unfairly criticised as bad when the full truth was exhaustion is not bad. As much as possible we had to practise a policy where all people were considered equal since the full truth is, as is now revealed, we were all equally good soldiers for humanity. Exhaustion or lack of it was not a sign of either goodness or badness but while we were insecure we misunderstood this. While we were insecure the innocent criticised the exhausted, seeing them as bad, and they retaliated against the innocent because they saw the criticism as unfair. Until we understood completely and were secure in that knowledge, our interpretations would be prejudiced by the insecurity of our understanding. While races do differ in their degree of tenacity and exhaustion, racist notions of groups of people being inferior or superior to others have no credibility. The full truth is, all people have fought equally hard <u>for humanity</u>, the only difference being some were more embattled from fighting <u>for humanity</u> than others. There is no such thing as inferior or superior (good or bad) humans.

It was mentioned earlier that instincts were simplistic. Trying to develop control using instincts required that the animal make every decision without reference to any general understanding or universal law. Instincts were not insightful although they were capable of guiding insightful thought. While the nerve pathways could be organised to ward against misadventures of the mind in the same way as reflexes were developed, natural selection could not select mental information associations (insights or reasonings or mental correlations) because these were not born with the animal. Insights could favourably or adversely affect an animal's chances of reproducing, and thus influence the genetic or innate makeup of the species but the insights themselves could not be selected, at least not directly. However, they could be genetically directed or channelled. Insights which improved an animal's chances of reproducing would be backed up or reinforced genetically because of this effect on reproduction capability, the result being that the insights themselves were also selected, although indirectly. Genetically we could not learn understandings, but we could learn orientations to understandings. While thoughts or mental reasonings could not be selected there was nothing stopping the nerve net being genetically organised (as occurred in the refinement of reflexes) so as to favourably incline the passage of certain thoughts towards certain associations/insights/ideas/reasonings, or even away from them. In this way our alienation or mental blocking out, our mental detouring or evasion of certain associations/insights, is now partially genetic in us. This is why, while the truth has always been fairly obvious, as is now revealed, the millions of humans on earth all found it fairly easy not to see it.

It was possible to let some truth through these instinctive blocks or evasions by means of prayer or meditation. By quietening our mind right down some truthful conscience guidance from our soul might struggle through the blockages and overburdenings to surface consciousness. Even Christ with all his soundness, his lack of alienation, at times had to employ fasting to make full contact with his soul and thus know what it knew. Denying our brain sustenance by fasting helped break up some of these instinctive evasions or blocks that are now in all our minds. One of the problems this book faces is that it says things that we have been blocking out for so long we will find it hard to hold them in our mind. No matter how clearly they are explained they will tend to drop out of our mind as soon as we stop read-

Prayer & meditation

ing. Even reading the book will be difficult for the more exhausted among us. The meanings will be blocked from coming through in our mind and we will think the information has not been clearly explained or even that nothing has been said. Christ experienced this when he talked unevasively. He said, using the metaphysical and highly critical language of his day, 'Why is my language not clear to you? Because you are unable to hear what I say . . . The reason you do not hear is that you do not belong to God [you evade integrative meaning]' (John 8:43-47).

Alienation is a very real phenomenon although, unable to defend it, we have hardly been game to mention it, let alone study it.

Humanity's insecurity

Instead we have coped by repressing the fact that it even existed. For example, while such things as 'human nature' and 'human affairs' are defined in most dictionaries, 'Human Condition' is never mentioned. That is how evasive we have been. The truth is, thinking could become so hurtful when we became exhausted that sometimes only constant chanting of mantras or counting of beads could stop the pain in our brain. In the decoration of some Middle Eastern mosques only motifs that have no resemblance to anything natural are used. The shape of a leaf would start the mind thinking of plants, then nature, leading inevitably to the confrontation with our 'unnatural' self that until now we have been unable to defend. Only alien patterns and objects, soothing cool colours such as blue and white and purging running water in fountains decorate the interiors. A great deal of modern Western interior decoration has taken this path, indicating a desperate need for relief emerging in the West. It has to be desperate because such alien decoration presents a false, unnatural view of our world, producing more falseness in those practising the escape and forcing others who are not already false to become so by coercing them into compromising their conscience. These examples illustrate how real alienation has been. Young people who have not yet adopted evasion will find the ideas in this book fairly simple and obvious while the older and more alienated will find it almost impossible to read.

(We can also see here the benefit of a slow mind and the danger of a fast one. A mind not very quick in understanding would still get there in the end and the preciousness of such slow progress was that the thinking didn't become lost — there was time for the conscience to have its say and guide the thinking. Instincts need time to express themselves. Watch animals that are not under threat make decisions and it seems to take them ages. A fast mind easily | **Danger of I.Q.** rides roughshod over its instincts, in which case, without its guidance system [conscience] our mind could only arrive at *The View From Nowhere* [to quote the title of Thomas Nagal's book mentioned at the very beginning of this book], which is precisely the view humanity had arrived at. Cleverness tended to know everything but in truth know nothing. While academia was quick to recognise the advantages of a high I.Q. it failed to appreciate the preciousness of a low I.Q. If academia had learnt to wait for slow minds it would have been amazed at what these slower minds could tell it/explain when they eventually caught up. We ignored/evaded/denied and were thus unaware of the thinking powers of introspection or soundness. For example, it was said of Christ, 'how did this man get such learning without having studied?' John 7:15. The nature and roles of the mechanistic and holistic approaches to inquiry will be explained in detail in part 3 of this book however it might be mentioned here why this book does not have a bibliography. The few references used in writing this book were able to be incorporated in the text. The synthesis of explanation in this book was found mostly by introspection not research. Mechanistic inquiry advanced step by proven step while holistic inquiry advanced by reference to conscience, by conscience guidance. Conscience indicated what information was to be trusted and what ideas were to be held onto.)

Certainly a mind could be directed by its instincts. If our body were starving instincts would soon have our mind thinking about obtaining food. Insights could be and to a degree were channelled instinctively. Such genetically influenced/directed thinking did, in its effect, represent innate insight or innate knowledge. Instincts have had a big say in the way we thought. We have lots of innate knowledge in us.

The *Penguin Dictionary of Biology* describes instincts as 'elaborate reflexes' and really that is true. Reflexes led to instincts which in turn lead to reasoning. At each level the genetic component or influence lessened while the information-associating or brain refinement became more elaborate. Reflex, instinct and reasoning were different names for different degrees of the one thing, information association, rather than separate names for different activities, as we sometimes evasively chose to believe. In our insecurity, in order to make ourselves feel unique, we humans sometimes tried to reserve the power of reason for ourselves, attributing only reflexes and instincts to other organisms.

In most animal instincts the relating of events involved is fairly simple and in no way approaches the minimum level required for what we call 'reasoning'. By reasoning is meant the capacity to understand the relationship of events sufficiently to manage them, that is, manipulate or change them, first over the short term and then the long term. A seagull chick associating the red spot on its mother's bill with feeding is associating or relating information but not on a very high level. It is certainly not making the sense of experience that chimpanzees do when they reason (associate disparate information) that by stacking up boxes they can reach bananas suspended from the top of their cage. Such perception or cognition, such consciousness, can only be achieved once a species has breached an impasse that appeared in the development of thinking capability. This impasse will now be explained.

3(d) Instincts Encountered the Limitations of Genetic Refinement

Through the imposition of instinctive constraints on its self-destructive capability, the mind was allowed to develop. This was all very well up to a point, the point where genetic refinement's limitations to learning integration were encountered. At this point all the limitations of genetic refinement began to limit brain refinement too.

Genetic refinement had three limitations:

1 It could not reinforce exclusive specialisation or division of labour because the reproductive unit would not be retained (except between the sexes and where there could be elaboration of the reproductive unit).

2 It could not reinforce altruism or love or pure (unconditional) selflessness because the sacrificed trait would not reproduce and so would not get an opportunity to continue into subsequent generations.

3 It could not reinforce a change which was an anticipation of the future unless that change first happened to suit the present (that is, conferred a reproductive advantage in the present). Only then could it become reinforced genetically, instinctively. Genetic refinement only 'commented on' the effect of changes in the present, it was unable to learn about change itself.

Genetic refinement's preoccupation with the present meant that it was only by chance that what a species learnt for the present also suited the future. With the advent of brain refinement, where the genes followed and in effect 'watched over' the brain, this situation was changed. The brain had insights about the future but the genes were only affected by how well changes brought about as a result of those insights met the present needs of the species. Instead of the present happening to suit the future, the future, in the form of the anticipation, had to suit the present.

Genes or instincts could not reinforce an understanding which resulted in changed behaviour if it did not meet the genetic need to maintain the opportunity to reproduce. Obviously, behaving unconditionally selflessly did not do this, so the instincts blocked the mind from reaching the understanding of unconditional selflessness which was fundamentally important to integration. Genetic refinement stopped the mind thinking anything which amounted, in the present, to truly selfless behaviour.

The effect was to forbid the mind to relate information beyond a superficial level. The mind was not permitted to appreciate (understand or relate or deduce) the fundamental theme in development, which is selflessness or love. Effectively the mind was being forced to stay stupid, unable to begin to make any real sense of experience. The situation was like having to play a game of football without being allowed to learn that the object was to score goals. In such circumstances the result would be a confused collection of players wandering aimlessly about the field. Animals trying to develop understanding, to make sense of experience, could not do so. Like the players on the field, they were faced with wandering around aimlessly with confused expressions on their faces and only their instinctive orientation to guide them a little.

This blinkering effect of genetic refinement meant animals could not learn to effectively understand (the relationship of events that occurred through time), could not learn integrative meaning. It was possible to instinctively encourage integrative thinking up to the level of reciprocity where selflessness is accompanied by selfishness — where others are given something in order that they will give something back. However integrative thoughts that did not lead in the majority of instances to the reproduction of the member who thought them (thoughts that in the final analysis were not selfish) could not be encouraged; indeed they were instinctively discouraged. Selfless/altruistic/integrative thinking, truly effective thinking, was blocked by the instincts just as surely as young animals were instinctively blocked from walking over a cliff.

Despite the apparent evidence to the contrary, love or unconditional selflessness, consideration for the larger whole, characterises the whole of existence. It is the theme on earth. It is what meaningfulness is. Just as you could hardly begin to integrate if you could not learn selflessness, so a cognitive system (a system that could understand the relationship of events through time) could hardly begin to appreciate integration if it were not allowed to appreciate the significance of pure

> **Explains why non-primates are stupid**

selflessness. Being denied this appreciation, the mind could not begin to make sense of experience, it was being forced to stay stupid; it was stalled at the level of superficial understanding. This is the impasse facing the majority of animals today. If we look into the eyes of a cow or a cat we see a confused mind kept stupid by this impasse, not conscious of its full relationship with events through time. If domestic cattle could relate information effectively, could think or reason, they would not graze so contentedly as they would likely deduce that market day and the butcher awaited them.

Just as we have become alienated from the truth and unable to think straight or be unevasive, so the mind in its initial development was alienated from the truth and rendered incapable of straight thinking. The human mind has been alienated twice in its history: once prior to its/our infancy (which was prior to the emergence of consciousness), then again during its/our adolescence (during the search for meaning), which we have just lived through.

By denying the developing mind the opportunity to relate information beyond an elementary level, by reinforcing only selfishness and not selflessness, the instincts effectively started to 'lie' to the mind, with the result that it could not develop any further, could not become more sophisticated in relating information. This brought an end to the development of understanding of change and thus anticipation of change in all species. Consciousness, awareness of the true relationship between events through time, was blocked. For instance, self-awareness or consciousness of self, which depended on understanding at least the relationship of the self with immediate events, was blocked.

Instinctive restraints had the double-edged effect of both making it possible for the mind to appear, which was the first stage in the mind's development, and of hobbling it once it had appeared. What was the second stage in the development of brain refinement which freed it from the restraints imposed by the instincts?

3(e) Second Solution: Love-Indoctrination which was Humanity's Infancy

While love-indoctrination has already been introduced it will now be placed in its context within the story of development.

We mentioned earlier the example of a mother bear being capable of learning genetically to sacrifice her life to protect her young, which increased the opportunity for the continued reproduction of her genes. The behaviour is called maternalism. It is in mater-

> **Explains the prime-mover in human development**

nalism that the opportunity existed to learn complete integration. While the maternal sacrifice cited is a case of reciprocity or gene selfishness it is also, in outward appearance, completely selfless behaviour.

The following analogy will help to explain. An oak tree and its acorn (seed) are two separate, although related, individuals. Behaviourally the 'mother' tree is altruistically or selflessly looking after the young acorn even though genetically the tree is acting selfishly by fostering its genes in its offspring. Imagine that the acorn had a self-learning device, a brain. During its time attached to its apparently selfless 'mother' tree, what would it learn? It would learn pure altruism or integrativeness or how to be utterly co-operative or selfless. It would be trained or indoctrinated in love, a process which throughout this book has been termed 'love-indoctrination'. Further, extending the period the acorn spends on the tree before it is shed would increase the amount of love-indoctrination it would receive.

Of course it was not in a plant but in animals that the self-learning device, the brain, existed, making the process of love-indoctrination possible. It is babies who, while they are completely dependent on their mother's apparent selflessness or generosity or love, are being indoctrinated in that selflessness or integrativeness. Having been so trained in infancy they will then behave integratively as adults. The longer they can be kept in infancy the more infants can be love-indoctrinated, the more they will practice selflessness or love or co-operation as adults and the more integrated groups of large multicellular animals will appear. After a while the 'genes following the brain' will reinforce this process and make love an instinctive expectation.

The 'trick' in love-indoctrination was that while maternalism was genetically selfish and therefore genetically encourageable, from an observing mind's point of view it was selfless behaviour. This loophole made it possible to encourage the development of understanding in the face of instinctive blocks that would otherwise have prevented it.

How could selfish genes reinforce a non-selfish process? Primarily by encouraging maternalism itself. Maternalism became much more than mothers protecting their young, it became a case of mothers actively loving their infants. In recognition of this we now talk of 'mother's love' not 'mother's protection'. Maternalism was genetically selfish but it trained a brain in selflessness. Once individuals appeared who were trained in selflessness that selfless behaviour would become reinforced by the genes. The genes would follow the training in love (the love-indoctrination) reinforcing it. While at times an expression of selflessness could still mean elimination of the individual involved, selfless behaviour was now appearing in the species in spite of such losses. Similarly, when the mind later went its own exhausted way during humanity's adolescence, the genes would follow along behind, as it were, 'reinforcing' what was happening. Generations of humans whose genetic makeup in some way or other helped them cope were selected naturally, making our exhaustions somewhat instinctive in us today. We have been 'bred' to survive the pressures of the human condition. The genes would always naturally follow and reinforce any development process, in this they were not selective. The difficulty was in getting developments to occur, not in making them instinctive because that was automatic. If it were not for the ability of the genetically selfish trait of maternalism to train the minds of individuals in selflessness, selflessness would not have occurred to be reinforced genetically because selflessness is a self-eliminating trait. The genetically encourageable selfish trait of maternalism made it possible for selfless behaviour to occur and develop in the species.

As love-indoctrination began to be developed genetically, the mind also began to support the process. We self-selected integrative traits by seeking mates who were loving. These mates were those members of the group who had spent a long time in infancy and who were closer to their memory of infancy (that is, younger). The older we became the more our infancy training in love wore off. We began to recognise this — to recognise that the younger an individual was the more integrative he or she was likely to be. We began to idolise, foster and select youthfulness because of youthfulness's association with integrativeness. The effect, over many thousands of generations, was to retard our mental and physical development so that we stayed as infants when adult. We became infant-like adults. This explains how we came to regard neotenous features of large eyes, dome forehead and snub nose as being beautiful.

> **Explains our concept of Beauty**

It also tells us why we lost our body hair and when this loss occurred. The physical effect was that we became infant-looking compared to our adult ape ancestors. We selected for what we now recognise as innocence. (Although later, during humanity's adolescence, we would become upset and resentful of innocence and instead of cultivating it would seek to destroy it. The attraction of innocence to mate with became perverted. This perversion of the act of procreation that we now refer to as 'sex' has already been explained.) Incidentally it was bullshit (our colloquial term for evasion; swearing was often a way of 'letting fly' — of being honest) to claim, as some have, that cuteness developed in order to look helpless and thus evoke sympathy. We knew that nurturing was all about loving but we have had to evade this truth. We often said 'babies are so lovable'. The truth was in the words we uttered even though we went on to deny it.

> **Explains our loss of body hair**

There is a marvellous illustration of the development of neoteny in an article that appeared in the April 1982 edition of the *Smithsonian* magazine titled *Livestock-guarding dogs that wear sheep's clothing*. The authors, Lorna and Raymond Coppinger, 'believe

the many breeds of domesticated dogs are derived from their
common ancestral wild type by neoteny — retarding development at some juvenile stage'. The authors divide the maturation
of a puppy into four stages. The first stage is characterised by
such behaviour as the puppy licking its mother's face to stimu-
late food delivery, some fighting over spoils with litter mates and
the tendency to scurry for the den yelping if threatened. Second
stage pups play with objects. The third stage is characterised by
'stalking' behaviour, pouncing and short chases to cut ('head') off
a litter mate's retreat. In the fourth, pre-adult, stage the pups
start following a parent ('heeling') and may even participate in a
hunt. The authors argue that cattle driving dogs or heelers such
as Welsh Corgis and Australian Blue Heelers have had their
mental and anatomical development retarded at the fourth
stage. For instance they have the pricked ears characteristic of
this stage in wild dogs. Collies that muster or round up sheep belong to the third 'heading' stage and have the characteristic half
pricked or 'tulip' ears. Most pet breeds fall into the second stage:
flop-eared, broad-headed, object players, chasers of sticks and
balls. Hounds, retrievers and spaniels are retarded or 'stuck-in'
this stage. Shaggy 'livestock [sheep]-guarding' dogs that stay
with the flock day and night to protect them from predators are
of the first type. They have the looks of fluffy puppies. They play
with each other and ignore sticks and balls. They lick the faces of
the sheep and their behaviour towards the sheep are the
responses of a puppy in loose association [integration] with the
rest of its litter. Their apparent aggressiveness — their barking
— is also characteristic of this first stage. The article says 'that in
a relatively short period of time, perhaps as little as 10,000 years,
the dog has adopted many shapes. Breeders continue to change
these shapes and behaviour by speeding up or slowing down
(retarding) the developmental rate.'

To reveal how important self-selection
was in human development what has been
said above about the speed of the development of breeds of dogs can be compared

Explains speed of human development

with a statement made by Jacob Bronowski in the book, *The Ascent of Man* (1973), which accompanied his TV series of the same name: 'We have to explain the speed of human evolution over a matter of one, three, let us say five million years at most,' Bronowski stated. 'That is terribly fast. Natural selection simply does not act as fast as that on animal species. We, the hominids, must have supplied a form of selection of our own; and the obvious choice is sexual [mate] selection.'

So love-indoctrination's earliest achievement was the establishment of the first totally integrated group of large multicellular organisms on earth — Childman. While humans became upset and divisive during humanity's adolescence, we lived utterly integrated lives during humanity's childhood, which lasted from five million years ago to two million years ago . As it says in the Bible, we were once 'the image of God' (Gen 1:27) and in Christ's words: it was the time when we 'saw God's glory' — a time when 'God loved [us] before the creation of the [hurt] world' (John 17:24) and 'the glory before the [hurt] world began' (John 17:5). Also the time when 'God made mankind upright [uncorrupted] [before] men went in search of many schemes [understandings]' (Eccl 7:29). Being trained in love as youngsters we practised it as adults, we co-operated with each other. We considered the importance of the group above the importance of ourselves. That was love-indoctrination's first achievement.

Now to introduce its second achievement.

Love-indoctrination had one remarkable side-effect. It liberated the brain from instinctive blocks, permitting it to think clearly. With love-indoctrination occurring, truly selfless integrative thinking was at last being promoted; the mind was free to think properly, soundly, effectively and so become conscious (of the real relationship of events through time). Consciousness is the essential characteristic of mental infancy.

How did love-indoctrination overcome the instinctive blocks? What actually took place?

At the outset the brain was small; individuals had only a small amount of cortex (where information is associated) and therefore could only relate information at an elementary degree. At the same time there were instinctively installed blocks or shepherdings orientating the mind away from any deep or meaningful perceptions. At this stage these small minds were then love-indoctrinated (trained in or taught selflessness or integrativeness during infancy). So although there may have been little unfilled cortex available what there was was being in-

Emergence of Consciousness

scribed with an effective information associating network of pathways. The mind was being taught the truth, being given the opportunity to think clearly, on top of and in spite of the already installed instinctive 'lies' or blocks. At first, with the brain so small, this truthful 'wiring', this imprinting or channelling, would not have been very significant but gradually it could be developed.

If the mind had not been given this training, integrative or meaningful thoughts would not have occurred because of the blocks (against self-eliminating behaviour) that were in place. On its own the mind could not have thought meaningfully/truthfully. Such truthfulness had to be forced upon it. Lions do not naturally jump through fire but in spite of this instinctive resistance can be taught to do so by circus trainers. We are able to, and often do change or mould our own and other animals' natural/instinctive behaviour. Many animals are unavoidably trained in love/integrativeness when they are young with their mothers and brothers and sisters. When they grow up and are forced to live alone (mostly because of the limited availability of their food source) as happens in many species, tigers and orang-utans for instance, they find it very hard to be thrown out of this love. We term this traumatic adjustment period weaning. In these situations the species did not want the integrative training but could not avoid it. The point to realise here is that the brain could be trained with love if something were there to train it — and love-indoctrination was there to train it.

So the mind was trained or 'brain-washed' with the ability to think in spite of the genetic blocks already installed to prevent such thinking. Of course it must be remembered that the emphasis in this early stage of the development of love-indoctrination was on the training in love not on the liberation of the ability to think — that was incidental to the need for a co-operative group member which love-indoctrination could produce. A species developing love-indoctrination needed a better love-trainable brain. It was coincidental that this love-imprinted mind also had the power to begin to reason or think or associate information effectively and consequently had the potential to eventually become conscious (of the true relationship of events through time).

The development of thought, which had been liberated accidentally, occurred only gradually. The requirements for a brain that could be trained in love weren't the same as the requirements for a mind that could think. The great associating cortex of our brain didn't develop strongly until thinking became a necessity in humanity's adolescence (when we had to find understanding to defend ourselves against ignorance). Intelligence was an asset to Infantman and Childman but not a necessity.

Though the early mind was not driven or needing to think, simply by being able to it gradually started to. Chimpanzees, which are in mid-infancy today, are beginning to become conscious of the true relationship of events — are beginning to be able to think straight (as evidenced by their ability to effectively associate information, such as reasoning that by stacking boxes one on top of another it is possible to reach bananas tied to the roof of the cage). The real value, so far, of love-indoctrination to chimpanzees has been in producing the close co-operation needed for their survival. (This need for love-training for survival will be explained shortly.)

Love-indoctrination's immediate value was in producing co-operatively trained members for the group. Only gradually did the liberated thinking power develop and only much later, when the mind started to try to self-manage, did the instinctive integrative orientation or conscience that love-indoctrination eventually produced become of value in guiding the development of the mind's understanding.

In order for love-indoctrination to develop there were three essential requirements. The first was that the species have the ability to look after a long-infancied, helpless infant. (In fact, to ascertain the level of intelligence of any species, we have only to find out two things: the species' need to develop co-operation and its ability to look after a helpless infant. For instance, when calving, whales leave their polar feeding waters and travel all the way to protected inlets in warm water latitudes where they experience starving conditions. The benefit of this practice is that it allows the calves to remain in infancy as long as possible and thus develop some love-indoctrination and thus some co-operativeness. Consequently but coincidentally the whale mind has also been liberated to some extent. If whales stayed in their polar feeding grounds to breed the danger the calves would face from exposure to the harsh elements and predators would mean the calves would have to grow up much faster than they have to in protected warm water inlets. These bays are often called 'nursery bays' in recognition of the fact that whales use them for nurseries. The dictionary definition for nursery is 'any place in which something is bred, nourished, or fostered' and 'any situation serving to foster something'. The implication is that there is more going on than just the protection of infants. The infants are being fostered, loved. Zebra foals have to be up on their legs and capable of independent flight from a predator almost as soon as they are born. Consequently the zebra species can develop little love-indoctrination, little co-operativeness and little conscious thought.)

However nothing compares with the freedom of arms when looking after a helpless infant. These limbs, the province of primates, made it possible to keep infants in infancy for an extended period thus allowing for

> **Explains when we learnt to walk upright**

greater love-indoctrination, which is why the primates have been able to develop so much intelligence. Due to their arboreal heritage of armswinging, semi-upright movement through trees, their upper limbs were already partially freed from walking. This was the crucial factor in mind development. It means our ancestors would have learnt to walk upright early in this, humanity's infancy period, because they had to use their arms to hold the increasingly late-maturing, helpless baby.

As well as the exceptional facility to look after a helpless infant, the exceptional need to develop co-operation was also required for mind development. A species had to need to be co-operative/integrative for the opportunity to develop love-indoctrination to be taken up. Not pushed by this necessity, gorillas remain comfortably hidden away in the jungle living on abundant giant celery and the other great apes are living relatively safely in forests. The exceptional need for co-operation occurred in a tree-living monkey-like primate (probably very similar to a fossil ape called Ramapithecus) some twelve million years ago. The cooling

Explains why other primates have not developed

world climate at this time was causing the forests to shrink, forcing our Ramapithecus-like ape ancestor to abandon life in the trees and adapt to conditions on the African savannah. Unable to run fast and without sharp teeth to defend itself as other savannah-adapted species could, this ancestor of ours was forced to depend upon and therefore develop co-operation between individuals as its main means of defence and survival.

Lastly, for the process of love-indoctrination to be carried to completion, it had to be easy to leave a baby in infancy, ideal nursery conditions were required: comfortable climate and abundant food supply. (Obviously the nursery conditions couldn't be entirely comfortable since the dangers needed to

Humanity was matriarchal

require co-operation between individuals had also to exist. They had to be comfortable in every respect except for this external danger.) Only one part of our ape ancestor's probable range appears to have provided such luxurious conditions — the aptly described 'Cradle of Mankind' — the Rift Valley of Africa, near the equator. At this early stage in humanity's lifetime, nurturing the infants was the priority concern and so throughout humanity's infancy and childhood, from twelve million years to two million years ago, humanity was matriarchal. In this sense, women created humanity. However patriarchy was just around the corner, when humanity entered the agony of adolescence.

These maturation periods of infancy, childhood and adolescence are the evasive descriptions we have used to describe the stages of the emergence of brain understanding capability or self-realisation — specifically of self-awareness or consciousness of self (infancy) followed by self-confidence (childhood) followed by self-understanding or identity search

| Humanity's stages of maturation |

(adolescence). They were to be followed by the adult stages of self-implementation, self-fulfilment and self-maturation. Once the infancy stage is completed in a species the other stages will naturally follow, although the rate of development may vary.

Infancy is the establishment of self-awareness or self-consciousness — of 'I exist' or 'I am' or 'I'. This discovery of the 'I' at the centre of changing experiences was the first major result of the emergence of the ability to relate information sensibly. Prior to the advent of love-indoctrination, the landscape of understanding (of change of the relationship of events) surrounding the animal was just a blur with the mind unable to make sense of the swirling array of experiences in it. This blur had to clear a great deal for the animal's brain (or mind or association or understanding capacity) to reason out its immediate relationship to events surrounding it, allowing the animal to effectively manipulate events, at least over the short term.

Prior to love-indoctrination the brain was a confused idiot. Compare the eyes of a cow or cat with those of a human baby. In the former you see this frustrated inability to relate — to understand — to think clearly; from the baby's eyes you can see the infant is conscious or aware of its world. The infant thinks and the thinking is starting to make sense to it. It is starting to register the relationship of events surrounding it. It is starting to understand.

The ability to learn to relate events beyond the immediate, over the much longer term — ultimately for all time — was to develop later on. These first understandings of immediate change meant the animal was recognising its relationship to its immediate changing surroundings. This emergence of self-consciousness, this awareness or understanding of self in the world, happened during humanity's infancy. This capacity made it possible for the animal to effectively manipulate events in the short term, such as our chimp reasoning how it can reach the bananas and human babies who now learn that by hitting a rattle they can cause a noise.

Gorillas and chimpanzees have been taught to 'talk' with sign language but while they reveal an awareness of identity or recognition of themselves in the world, their level of language development is only that of an infant. W.H. Thorpe, in an *Encyclopedia Britannica* entry, *Learning, Animal*, discusses the sign language used by Washoe, a female chimpanzee, with Roger, her trainer. Washoe could sign 'you Roger me Washoe go out'. 'After four years of study,' Thorpe says of this, 'the chimpanzee (Washoe) seemed unable to use signs for such combining words as "and", "for", "with", or "to". It is significant, however, that words of this type are noticeably absent in the early sentences of young children. Hence, although Washoe's achievements are not those of a fully articulate adult human her earliest two-sign combinations were comparable to the earliest two-word combinations of children.' Chimpanzees as a species are in mental infancy or are equivalent to infant humanity.

The ability to self-adjust (a primitive nervous system such as exists in planaria worms is capable of self-adjustment) is different to self-management (of which humans are capable) and the insecure self-management which humans have been practising is different to the secure self-management we will be capable of now we have found understanding. Self-management, the ability to manage events, emerged with consciousness. However, it is one thing to control a rattle — to manage immediate events — but quite another to properly manage or control the destiny of the earth — to manage events over the long term — to be secure in our management of ourselves and our world. Before we could be secure competent managers we needed to know the meaning of existence and be able to conform to it.

In adolescence we faced the responsibility of having control of events over our entire lifetime and the lifetime or destiny of our planet. This did not mean that we exercised meaningful secure control, merely that we had the ability to control. We had gained sufficient understanding of the relationship of events to manipulate them but not enough to manipulate them securely, which required a complete understanding of existence. The difference is between being able to do whatever we like and being able to do whatever is meaningful. This transition from free will to responsible free will was the agonising journey of adolescence.

In humanity's infancy the newly acquired ability to relate events, the liberation of understanding from instinctive contradictions, conferred the freedom to practise doing so. Our infancy saw the | **Explains free will** |
first displays of mental independence, the first displays of consciousness, reason, free will.

Despite these displays, the infant was still totally dependent on its instinctive orientation or knowhow for safe passage through life. Infantman's developing integrative instincts or love-training or mother equivalent looked after Infantman. From that safe shelter Infantman was able to experiment a little on her own without endangering her life — was able to begin to play — was able to begin to pretend to be an adult. Free will in infancy mostly manifested itself in a realisation of self; it marked the emergence of consciousness or awareness of self. It was not until childhood that all-out implementation of mental independence or playing in earnest began.

(Note 'her' and not 'his' is used because humanity's infancy and childhood was matriarchal or female-role prominent, so that 'her' is slightly more appropriate. Childman and not child-woman is used as it is our abbreviation for childhumanity. 'They' instead of 'he' or 'she' could have been used however, as has been explained, we are now unevasively talking of the species as an entity, so the singular pronoun is more appropriate.)

To summarise to this point. Love-indoctrination's earliest achievement was the development of the first totally integrated specie society or group of larger multicellular animals. Its second achievement was the liberation of the mind to think or relate information effectively.

Now its third achievement has to be introduced. This was that it correctly orientated the mind. Love-indoctrination gave us our conscience.

As explained already it was maternalism, the behaviour that trained an infant in selflessness, that was directly genetically encourageable. Women acquired what we now refer to as their 'strong maternal instinct' during the period of humanity's development when maternalism was being encouraged genetically. These strong maternal instincts were more than just for the need to look after helpless infants as we have evasively claimed, they were for training the minds of our infants in love. Being able to encourage love by encouraging maternalism opened the way to developing instinctive reinforcement for love. A loophole in the process of refining genetically had been found: an instinctive trait that was selfish and thus genetically encourageable but which produced love or pure selflessness. All that was left was for the whole process to become instinctively reinforced — as was now possible — and love or integrativeness would become an instinctive expectation in us. That happened. The result was our integratively orientated conscience.

Love-indoctrination nurtured a mind into existence, liberated thinking and orientated that thinking. It gave us both consciousness and our conscience. It has been the most powerful process in all humanity's development.

Our conscience would be vital during humanity's adolescence. It was our conscience which correctly orientated the mind so that when it set out in search of understanding, it had a perfectly accurate guidance mechanism to help it and make the search possible. Of course a great tragedy — the greatest tragedy humanity has known — lay in the way our conscience guided us. When we made mistakes in self-management our conscience implied that we were bad. It had no sympathy for our necessary search for understanding. While we needed our conscience's guidance we didn't like the way it gave it. We didn't like having to live with an unjust sense of guilt. We needed guidance but we did not deserve criticism. The point here however is that, upsetting as it was, we could never have found understanding without our conscience. If some other phenomenon had liberated the mind it could have started thinking but without an orientation towards integration could never have completed that thinking and found meaning. (Of course nothing but love or integration indoctrination could have liberated the mind but the dual effects of love-indoctrination become clear in this imagined situation.)

Without our conscience to guide us during our adolescent search for understanding we would have practised the inevitable initial misunderstandings of selfishness without constraint, the result being that we would soon have destroyed our society or group. Without its conscience, humanity would never have had the opportunity to find understanding, to find itself.

What this means is that, while it is true that if it weren't for the instinctive blocks to effective thinking the brain acquired long before it became purely human (or even primate), all animals could have developed consciousness, they could not have gone on to find understanding without first acquiring a conscience to guide their search.

The fourth and final achievement of the process of love-indoctrination was to free the hands to hold tools and implements. The more love-indoctrination developed, the longer infants were kept in infancy, the more they had to be held, the more upright we had to become in order to hold them. This releasing of the hands would prove vital later on when our mind needed to express itself because it provided a means for that expression. The mind could direct the hands and the hands could manipulate the world. (Interestingly, anthropologists called the first variety of Homo or intelligent man Homo habilis, which means 'handyman'.) A fully conscious mind on a whale would be frustrated through its inability to express itself; conversely, the free upper limbs of a kangaroo are useless without a conscious mind. But couple a conscious mind to hands and a very creative force emerges.

In all, love-indoctrination was a truly extraordinary opportunity for God or development and of all the species we humans were the ones sufficiently pre-adapted or suited to develop it.

Due to the development of love-indoctrination our ape ancestor, Infantman, genetically learnt completely integrative instinctive reinforcements. The great apes and baboons are currently about half-way through the infancy period of mental development. Recent primate studies, tainted by the precepts of sociobiology, seek to hide or evade the extraordinary integration present in these species because of the implicit criticism it makes of divisive humanity. Earlier primate studies, being more naive of this threat, revealed the integration. (Incidentally, in all areas of inquiry there has been an optimum period of enlightenment where we had gained significant knowledge but had yet to become too sophisticated [in evasion] or aware of the dangerous/hurtful/critical implications of that knowledge or insight.) Eugene Marais, who has already been introduced, was the first to undertake prolonged field studies of primates back in the 1920's. In his books, *The Soul of the Ape* (not published until 1969) and *My Friends the Baboons* (1939), he stressed just how integrative baboons are.

For instance, there is this passage from *My Friends the Baboons*: 'The females have certain rights which are scrupulously upheld by the whole troop. The males must protect and defend them in all circumstances. [This indicates how it was the women's role of nurturing that was the priority concern at this stage — how the situation was matriarchal.] If danger threatens, the full-grown males form a vanguard, and they will often sacrifice their lives to prevent an assault on the females and little ones [Note the male's role as group protectors]. The wife is always entitled to a share of the food her husband gathers.' Note the pure selflessness or altruism here. To quote the Bible, 'Greater love has no-one than that one lay down his life for his friends' (John 15:13). (Of course we must not forget the greater truth that adolescent humanity's human's preparedness to lose/corrupt themselves in order that the true and beautiful world might one day be permanently established on earth was also an act of unconditional self-sacrifice. While we often appeared selfish, the greater truth was we were being selfless and sacrificing our life for a greater good.) Dian Fossey's studies of the mountain gorilla (see her book *Gorillas in the Mist*, 1983), although in places tainted with sociobiology, also reveal the strong relationship between nurturing and integrativeness which is the process of love-indoctrination (although of course she was not aware of the significance of the process or even aware that what she was describing was a process.)

In terms of development, baboons and apes are where humanity was about eight million years ago. By five million years ago humanity had perfectly refined integrative instincts that were even more integrative than these just described for baboons. For instance, among both baboons and gorillas there is still a residual amount of as yet uncontained sexual opportunism, such as infanticide (where a newly dominant male will kill the offspring of his

| Infanticide |

predecessor, bringing the mother of the murdered infant back into season earlier than would otherwise occur, allowing the new dominant male to mate and reproduce his genes more frequently), and friction within the group between males competing for mating opportunities, creating the need for some dominance hierarchy.

3(f) Childhood

By five million years ago, as humanity left its infancy period, we had acquired perfectly refined integrative instincts. During infancy our indoctrination in love had become fully instinctively reinforced. Any thought and consequent act was completely integratively orientated or shepherded. Whenever the mind thought integratively the instincts in effect said 'that's right' and whenever it thought divisively they said 'that's wrong'.

It is important to remember though, that it was not because Infantman understood that the meaning underlying change was integration that she attempted to be as integrative as possible; it was because integration was needed as a means of survival. In other words, integrativeness was a genetically imposed orientation not an understanding or insight at this stage. We learnt <u>how</u> to be integrative but not <u>why</u> we should be integrative.

Humanity's infancy was a difficult period for development. It was not easy to develop and maintain love-indoctrination. It was like trying to swim upstream against a fast current to reach the safety of an island. Any disruption or lack of attention to the task of loving infants would produce divisive adults who could not practice the co-operation necessary to maintain the group. In trying to reach the safety of the island of co-operative behaviour Infantman often failed and was swept back downstream to the pre-love-indoctrination situation where such divisive, disintegrative behaviour as sexual opportunism reigned. But co-operation/integration was finally achieved, marking the end of this stage in development and removing the pressure for development.

With the pressure off, our ape ancestor began to multiply or flourish as she moved out of infancy into childhood. (Such increase occurs naturally following the completion or fulfilment of any opportunity for genetic development because survival stability or relative non-change

> **The reason for 'spasmodic evolution'**

returns to the species. Implicit here is the explanation for so-called 'punctuated equilibrium' or 'spasmodic evolution' that has confounded evasive science — if you are not going anywhere, as the idea of evolution maintains, you can't have an impasse or break through an impasse, which in truth was what was happening).

This population explosion marked the first major success for brain development and the transition of humanity from infancy to childhood. As a consequence of the population increase numerous fossils have been found of Childman (the australopithecines on the archaeological scale). Childman lived in what we instinctively 'remember' as paradise. We describe it in the Bible as 'The Garden of Eden' and in Australian Aboriginal mythology it is aptly called the 'dream-time'. This was

Paradise and The Garden of Eden

the time we finally became as God or development wanted us, integrative, in the 'image of God'. (Of course while development had at last achieved the complete integration of a large multicellular animal species, there was still the task ahead of producing a system capable of understanding change. A mind had still to be fully developed. We were in the 'image of God' but not yet understanding of God or 'knowing'.)

Humanity's three-million-year childhood, which began five million years ago, was spent idyllically. Until the appearance of naughtiness and some cruelty towards the end, our behaviour towards each other during our childhood was utterly integrative, giving and loving or unconditionally selfless. We lived in the shelter (on the apron strings) of our 'mother', love-indoctrination, 'who' looked after us, 'who' told us how to behave. This was the idyllic integrated world to which we can now at last return.

Childman was successful but she was under no particular pressure to develop in any new direction. The question then is, what led to the development of the mind through childhood into adolescence? We know that love-indoctrination itself did not promote a better mind it only liberated mental cleverness or

156

reasoning. However, this would have been of some survival assistance and therefore the development of some mind or association cortex would have occurred during infancy, such as to the degree of association exhibited today by the chimpanzee stacking up boxes to reach bananas.

The self-confidence that came with more intelligence would have led to some mind development during humanity's childhood through mate selection, because admiration for cleverness as an aspect of self-confidence in ourselves and in others would have been expressed in the choice of a mate. While there was no pressure to change or develop more intellect there was also nothing stopping the development of intellect. Childman simply wandered towards adolescence. It was like a flock of sheep wandering into a lush pastured field with the gate (into adolescence) open on the other side. While there was no need to go across and no reason not to, there was also nowhere else to go, so the sheep spread out in the field and eventually found their way to the gate. While development of the mind was not forced during childhood, nor was it restrained, so gradually it occurred.

Infancy was a period of inward orientation (it was about being loved which liberated and nurtured or established the conscious thinking self) while childhood was a period of outward orientation (it was about practising with the now liberated conscious thinking self and discovering its power). Infancy was about 'I am' and childhood was about 'I can'. Infancy was the development of self-awareness and childhood was the development of self-demonstration — of self-confidence — of what self (the conscious thinking self) could do. In childhood we began to apply our ability to make things happen. This was a time of reinforcement of the mental self, of drawing it out, and it was an exciting time. (Incidentally this description of our psychological stages of development shows how watching television can be so bad for today's children. Children have an intrinsic need to make and do things themselves, not watch other people making and doing things. They need an active reinforcing environment not a passive one like television. We take creativity — the power of the mind to manipulate events — so much for granted now because we have been living with it for so long that we can forget how amazing our ability to make, create and do things is. Childhood is the time we should naturally discover this amazing talent. If children discover the power of free will sufficiently well they will enter adolescence inspired to find understanding of this power. What children need to be properly prepared for adolescence and adulthood is reinforcement. 'Toughening' them or trying to make them egotistical/winners has no place in the life of a child. They will be naturally strong and winners later in their life if they are loved in infancy and reinforced in childhood.)

Although in its adolescence humanity would have to find understanding, in childhood our self-management mistakes had not yet got sufficiently out of hand to require this. Humanity's childhood was an idyllic time in which we were gloriously free of the responsibility and the consequences that came with free will, this wondrous ability to do whatever we liked with ourselves and our world. We only dabbled in self-management. We still depended on our 'mother', love-indoctrination instincts, for safe passage through life, but we were beginning to investigate the limits of our power to anticipate. We were hanging onto our mother's apron strings very firmly with one hand while using the other to conduct brief exercise 'games' in self-management, not because we had to, but because we could. We were 'playing' (with self-management) as we evasively have described it.

Humanity's early childhood can best be described as our 'prime of innocence' period. It was when we revelled in our emerging intellectual freedom in the perfectly secure shelter of our maternal training, unaware of the dangers of such freedom. It was the demonstrative stage 'look what I can do — I can jump puddles — aren't I fantastic'. It was the emergence of pride in self, in our mental or self-management capability. It was the emergence of mental confidence and independence of instinct. However, there still remained the vast difference between being capable or confident of management and being capable of secure management. Early Childman would have been as naive (of the consequences and thus responsibility of mental freedom) and as demonstrative or happy in this freedom as a child is today. In humanity's demonstrative stage of early childhood we proclaimed our mental independence.

By midway through childhood, some three and a half million years ago, the emerging self-management capability and independence from instinct would have started to get Childman into trouble. Innocent mistakes began to bring our self-management capability, our mind, into conflict with our perfectly integratively orientated instincts, our conscience. These

Naughtiness in children

earliest self-management misadventures which when displayed by children we call naughtiness characterised humanity's late childhood. As the word 'naughty' implies, they were not serious mistakes since at this stage our conscience could still repair and thus contain them.

The first mistake or misunderstanding was the obvious one of being selfish — of seeing ourself as being 'I' before seeing ourself as part of 'we' — of taking all the bananas instead of sharing them. It was the first time since pre-love-indoctrination times that selfishness had appeared. (It occurred again later, in humanity's adolescence, but these later acts of selfishness were for completely different reasons. We were angry and preoccupied demonstrating our worth and became divisive and self-centred and selfish because of our conscience's unfair criticism of our efforts to self-manage.) The consequences for Childman were devastating. Everyone else (other group members and her own conscience) was angered by her outright selfishness. Generally a child only made this grand mistake in understanding once before becoming cleverer or smarter and realising the benefits of a more subtle form of selfishness, reciprocal or conditional selfishness. This time the others were also given some bananas, to stop them becoming angry and/or in anticipation of sharing their bananas later. We may have pacified the other members of the group but our conscience was not deceived and became increasingly hurt by these misadventures in self-management.

From here on Childman became more and more devious in her childishly superficial rationalising of her 'if I can get it all for myself then why shouldn't I ?' attitude. (The reason such selfishness was 'superficial rationalising' or limited reasoning will be explained shortly.) Actually there were two forms of selfish opportunism possible in a group or association or system: greed and laziness; taking more than a share and contributing less than a share. While Childman had no reason to be opportunistic, initially she saw no reason not to be, either. In learning to understand God or integrativeness humanity had a long way to go.

The development of understanding had to go through all the stages of a system developing integration. A parasitic species characteristically starts by selfishly living off its host then, through natural selection, discovers this is ultimately a self-destructive attitude and so develops a relationship, called mutualism, which is a relationship of reciprocity. A parasitic relationship, like a selfish one, is an immature association. Because parasitic species learn by natural selection, they cannot develop altruistic capability which would promote completely integrative relationships, so development of associations of organisms or systems of this type stop at mutualism.

Biologist Ralph Buchsbaum (in his book, *Animals without Backbones*, 1948, in which he makes reference to parasitism in flatworms) says:

'Commensalism [where members of different species live in close association without much mutual influence] usually evolves, not in the direction of mutualism, but towards parasitism. A commensal that at first takes only shelter, then scraps of food, finally begins to feed on the tissues of the host body, and the host suffers a certain amount of harm. Should the parasite become so well adjusted that it causes little damage, and, in fact, finally proves to be some service to the host, the parasitism becomes a mutualism. Thus, the three kinds of relationship are only different stages in the process of living together. Mutualism and parasitism can probably arise directly from commensalism, but they may evolve from each other. Since mutualism requires the greater number of adjustments, it is relatively rare as compared with parasitism.'

It is easier to be selfish/parasitic but ultimately we learn that only co-operation/mutualism works in the long-term. This is a microcosm of all existence — of the development of integration. Even we humans, with all our built-in evasions, have been capable of revealing the truth in remote situations such as the life of parasites. The point to be made is that the stages in development of ecological systems and of the mind's understanding were, firstly, selfishness, which became reciprocity, which became pure integration (in systems capable of learning unconditional selflessness) which was the emergence of a larger, more stable system. Unconditional selflessness is the reason a human body works so well. Every cell in our body has submerged its individuality to the needs and functioning of the larger system which is our body.

Early Prime Of Innocence Childman was Australopithecus afarensis. Middle Demonstrative Childman was A. africanus and Late Naughty Provocative Bullying Childman was A. robustus. (Anthropologists recognise an early and late form of A. afarensis as well as an early and late form of A. robustus, the latter, slightly heavier form sometimes being referred to as A. boisei.)

The reason late Childman is described above as 'provocative and bullying' as well as 'naughty' is that by the end of childhood we were beginning to taunt/challenge the dangers of self-management to come out and show themselves. We have all seen children, boys especially, pulling off fly's wings, burning ants and pushing each other over in the playground. They become extremely naughty, even cruel. The reason was that their minds were saying 'why shouldn't I do whatever I want to do?' They were taunting their power, trying to discover its limits. They were throwing out a challenge (pathetic as it sometimes was) to the world to reveal to them the significance of this power. The reason boys especially tried to provoke this self-adjustment power to reveal its limitation (which was the need to have understanding), the reason they tried to draw out the danger or threat of ignorance that self-management contains, was that males have always had the job of group protectors. It was males who were going to be charged with the task of going out to challenge this group-threatening lack of answers/understanding — this threat of ignorance. Unaware, at this stage, of the magnitude of the dangers involved in this threat, boys have looked forward to and even provoked the impending battle.

What has just been said about the capacity of older children to be cruel can be better understood if we appreciate that cruelty to others is really only the other side of the selfishness coin. Greed and cruelty are the two extreme forms of divisiveness possible in our world. While our soul or instinctive self knew not to be selfish and not to be inconsiderate of or, in the extreme, cruel towards others, the first mysteries the emerging conscious mind would encounter was 'why not be selfish?' and 'why not be cruel?'. When we are older we take it for granted that cruelty is wrong. We forget that cruelty to others, like selfishness, had first to be 'tried out' by the mind before it can be rejected. They are the first two mistakes the mind makes in learning to understand.

Incidentally, the major physical difference between early Childman, A. afarensis, and late Childman, A. robustus and A. boisei, is that the latter had more pronounced cranial and facial bone structures. Anthropologists know that these modifications were for the support of much stronger facial muscles used to work the heavy jaw and huge grinding teeth of the later australopithecines. We know all the australopithecines were vegetarian but why did the later australopithecines need bigger grinding teeth? What change in diet occurred? And why? The answer lies in an appreciation of the different psychological state of the later australopithecines. They were more extrovert, increasingly naughty and roughly behaved. As well, like older children today, they would rather have been out playing than eating. To fuel this energetic robust lifestyle and let them 'eat and run' they would have needed a readily available food source that they could eat quickly. Being vegetarian, they would have needed a lot of it because vegetables are not as efficient an energy source as meat, for instance, which was not to appear on humanity's dining table for some time yet. (While australopithecines would have been capable of being cruel to animals they were not yet upset with animals and thus practising killing them regularly which was what finally led to meat eating.) We can imagine certain edible varieties of roots, tubers and stalks best filling this need for a ready fuel supply, which explains the need for massive grinding teeth and accompanying facial structure.

It is now possible to summarise the story of the emergence of brain refinement, the development tool which, together with integration and the other development tool, genetic refinement, completed the trinity of 'characters' involved in development. It was the advent of this last tool, the vast information-associating nerve network of the brain, that made it possible for humanity to learn to understand 'God' or Development.

The basis of nerve-based refinement was the ability of nerves to remember experiences. From this came the possibility of relating memories, of processing information in the nerves. When these related memories or insights became capable of effecting action through effector muscles, an organism was capable of altering its behaviour in anticipation of likely events. This marked a major turning point in development. Organisms were now capable of reacting individually to a possible future instead of reacting as a species to an existing present. Initially such anticipatory behaviour (nerve-based information refinement) was promoted by the species because it provided variety for the process of genetic refinement. This meant that only anticipations of the future which were valuable in the present were selected. Insights were shepherded and reinforced by the genes. These earliest orientations for insight-based changes in behaviour are what we term instincts. Instincts eventually developed to block further development of the mind's (the nerve centre that developed for this processing/comparing/relating of information) understanding or insight, its power to anticipate.

Consciousness or awareness of the deeper meaning behind changing events and thus the ability to anticipate the long-term goal of development depended on liberating the mind from these instinctive blocks. This was achieved during humanity's infancy by love-indoctrination. It was this ability to manage change or self-adapt (even if, during infancy, it was only for short periods) that would have promoted the mind's development through this period. Following this, during childhood, the mind's confidence in adapting its own behaviour emerged in the form of self-admiration, which played a part in mate selection and in this way promoted the development of the mind through this stage. Finally, during adolescence the understanding capability was needed and sought after to help relieve us from the unfair criticism levelled against us by our conscience and innocence in all its forms surrounding us.

Humanity's mind was brought to its full capacity for understanding stage by stage. It was a remarkable journey.

3(g) Adolescence and the Human Condition

During humanity's childhood our capacity to absorb and repair any hurt to our soul (caused by compromise to our instinctive expectations) was boundless. The balance of love and hurt was completely tipped towards love. We had so much to give that any hurt amongst us, such as a child in our group being orphaned and as a consequence missing out on some of his or her instinctive expectation of 'bonding' (love) and becoming upset, could easily be repaired. Our capacity to keep 'turning the other cheek', to absorb and repair upset was very high. This integrativeness/selflessness also made it possible to repair the 'mistakes' made in our first tentative experiments in self-management, the games we played in childhood simply because we could.

Finally, however, the number of 'mistakes' resulting from these games became too great for either our conscience to contain or our integrativeness to repair. Upset began to set in. The game suddenly became serious. This corruption breakout point signalled the end of humanity's childhood and the onset of adolescence and launched our desperate search for understanding because now only real answers could stop the upset increasing and spreading. The need to find understanding was suddenly imperative because it was the only thing that could stop the conflict between our conscience and our mind. We desperately needed to explain these acts our conscience saw as 'mistakes'.

Incidentally, it was this emergence of the need to explain ourselves that led to the development of language. Anthropological evidence supports this insight. Study of fossil skulls for the imprint left on the skull wall of Broca's area (the word-organising centre of the brain) suggests, according to Richard Leakey in his book *Origins* (1977), that, 'Homo had a greater need than the australopithecines for a rudimentary language'. Apart from contact calls our Childman ancestors had little need for sophisticated noise signals and no need to express concepts because their behaviour was instinctively co-ordinated and there was nothing to explain. This instinctive empathy we once had is now much repressed in us.

The advent of language

Nevertheless, we were once so in tune with each other and our world that, to quote Sir Laurens van der Post about the relatively innocent Bushmen of the Kalahari (from *The Lost World of the Kalahari*, 1958), 'He and they all participated so deeply of one another's being that the experience could almost be called mystical. For instance, he seemed to know what it actually felt like to be an elephant, a lizard, a baobab tree.' It has to be

How we lost our sensitivity

remembered that despite their relative innocence compared with the rest of us, the Bushmen are still members of modern highly embattled Sophisticatedman, Homo sapiens sapiens, so if this is the sensitivity we are capable of when living naturally (in the environment our soul is familiar with) today, how much more sensitive must Childman, who had not even engaged in the battle, have been! Her sensitivity would have been so great it would appear to us as supernatural. In two million years of battling and repressing our soul we have become extremely superficial and insensitive. We are almost completely numb beings. That this degree of repression has been necessary is a measure of just how much pain and hurt we have had to bear (since repressing hurt was our way of coping with it) and how tough and heroic we have been in the face of such pain.

There is a whole world of beauty from which, over two million years of repression, we have become almost totally alienated, from which we have lost almost — but not all — access. Our great artists — be they painters, sculptors, singers, musicians, dancers or poets — are essentially people in whom the alienation that now exists within all humanity is incomplete. Two million years of repression has achieved an almost but not quite complete block-out of our soul's awareness of the magic world of beauty on earth. Occasionally a human occurs whose protective block-out has a crack or tear in it. Through this small window these people can touch some of the beauty that exists on earth. Look closely at Claude Monet's paintings of water lillies. His amazing empathy/sensitivity is instinctive. Certainly he cultivated his talent but what this really means is that he cultivated ways to further let his soul reveal its immense sensitivity. In an interview with the Australian painter Martin Sharp about his art, radio talkshow interviewer Caroline Jones said 'are you saying that art makes visible things from another dimension' and Sharp replied, 'Yes, at times I seem to expand and do things gracefully instead of chaotically'. There was also mention of 'art supplying a rare glimpse of God'.[1] (By the way, this does not mean the innocence/soundness of prophets such as Christ is similarly due to a freak tear in their/our instinctive block-out. Unlike these exceptionally sound people many great artists had terrible childhoods. Their sensitivity arises from having a rent in what is now the basic fabric of instinctive alienation in humans, a peephole to beauty, that lack of nurturing wasn't able entirely to block-out/repress, although it did influence the way the talent expressed itself. People with such freak rents we call geniuses. Christ was not renown for any one particular freakish talent — he was not a genius rather he was just overall sound/innocent. Exceptional soundness has been so rare and our ability to cultivate it so denied that the best we could normally hope for in terms of talent was the appearance of these freak geniuses. This is a sad commentary on our plight.)

However, to return to Adolescentman's need to understand existence. We know that genetic refinement could separate information from matter (and thus adjust the species to present requirements) but not information from influence. This was only possible with brain refinement. The ability to process information separately from its influence made it possible to learn about change itself and thus to anticipate change; using our brain it was possible for us to adapt ourselves to the future, not just to present. However, to correctly anticipate the future, to become responsible/secure managers of existence, we had first to understand existence.

As said before, it was by chance that our ape ancestor, Infantman, needed to be integrative or co-operative. She was instinctively indoctrinated or trained in integrativeness or love because she needed it to survive but this did not give her an understanding of integration. This distinction — and conflict — between our instinctive integrative orientation, our conscience, on the one hand and our conscious experimenting self, our mind, on the other is important. It was a dichotomy which forced adoles-

---

[1] *Australian Radio National*, October 6, 1987.

cent humanity into serious self-management experiments in order to learn how to manage securely. No longer were self-management experiments a game of childhood.

The major conflict arose when the mind began trying to manage from a basis of limited understanding of existence and in the process made mistakes, such as late Childman's innocent decision to take all the bananas for herself. The instincts criticised these mistakes and tried to stop the experiments. We were in the impossible position where our conscience was trying to stop us doing the one thing we knew we had to — it was trying to stop us learning how to adjust yet we had to master this ability in order to meet the future successfully!

We were a self-adjusting system as yet unable to successfully self-adjust and dependent on our love-indoctrination, our instinctive shepherding, for guidance. Sooner or later we had to grow up and leave the safety of our infancy and childhood home to shoulder the responsibility of being a self-adjusting system, to achieve our **When we left Africa** potential. We set out on the adventure to find our own identity and thus learn to cope on our own. The first bands of early Homos (intelligent but insecure Adolescentman) left our ancestral home in Africa about one and a quarter million years ago.

The following stages of Adolescentman have already been introduced and explained in Step 5. The first Adolescentman who had to resign himself to the fate of having to self-adjust without answers/understandings was Soberedman, whom anthropologists know of as Homo habilis. Homo erectus followed, he was Adventurousman, the adolescent who left Africa. Then came Angryman, who was Homo sapiens (sometimes referred to as H. sapiens [archaic]) and finally ourselves, modern, battle-wearied/exhausted or refined in alienation Sophisticatedman, Homo sapiens sapiens (sometimes referred to as H. sapiens [modern]).

Adolescent humanity was in an ambivalent situation. Were it not for **Explains sin or evil** the restraining presence of our correctly orientating conscience we would have had no safe shelter or basis from which to conduct the search for understanding. For instance, it was the continued presence of the innocent — of our conscience, of idealism — within us and amongst us (contributed especially by each new innocent generation) which saved us from ourselves, kept us from excessive mistakes and the consequent alienation/corruption of our integratively orientated soul. In this way our conscience facilitated our search. Conversely, we had also to defy our conscience to some degree in order to master self-adjustment. We were in the extremely difficult predicament of having both to defy our conscience and obey it! This was the paradox within which we lived and struggled during adolescence. This is what modern society has called our Human Condition. Within this paradox is the origin of criticism — of 'sin' or 'guilt' — and also the origin of defiance of criticism, which is our human anger or aggression and resentment of criticism. We were in competition with the false implication that we were bad or evil. For instance, the presence of innocence only represented the unfair criticism we were trying to live with, so we resented innocence, we tried to evade its criticism, and we destroyed the innocent.

— We destroyed innocent animals (the advent of hunting and thus meat eating about one and three-quarter million years ago).

> **Explains hunting & meat-eating**

— In perverted sex we destroyed the innocence of women.

— Innocent people were replaced by less innocent people. Less intelligent, more innocent people could not cope in the new reality of compromise and struggle and were replaced by more intelligent, more realistic people. It was only fifty thousand years ago that development of I.Q. or speed of mental information processing came to a halt and a balance was struck between answer-finding but corrupting cleverness and non-answer-finding but sound lack of cleverness.

— We destroyed the innocent original self or soul within our-selves by mentally 'forgetting' it — by means of repression, block-out or evasion. The result was we became separated or split from our true situation and self which is

| Explains alienation and subconscious |

the origin of our alienation/psychosis/neurosis and also the origin of our subconscious (repressed) self.

The ideal solution to the paradoxical predicament of having both to defy our conscience and obey it was to have been balanced about it. If we obeyed our conscience completely we were safe for the present but would not learn to understand exis-tence; if we were excessively free of its influence we would make too many mistakes and diverge too far from integrativeness (would become too upset, which is angry/egotistical/alienated and superficial, all of which are selfish, socially disintegrative traits). If our society became disintegrated/destroyed we would never find the understanding that would end our upset. We needed to be sufficiently free to experiment but obedient enough to our conscience to avoid becoming 'corrupted' and socially dis-integrated. The trouble was such an ideal balance could not be known precisely so it could be found only through oscillating be-tween both extremes — through pursuing freedom until we be-come obviously excessively mistake-laden or 'corrupt', then changing to the pursuit of obedience to conscience until this, in turn, became too repressive of the need to understand, and so on, back and forth.

In politics, one side of the balance is represented by socialism and the left wing, where obedience to absolutes is stressed, the other by capitalism or the right wing which stresses freedom from absolutes. It is only the dialectic or oscillation between these excesses which reveals the middle ground. At the same time, these oscillations have given rise to

| Politics |

alternating bouts of excessive freedom and excessive oppression and thus conflict and argument and polarisation. We progressed by argument which, in the extreme, was war.

It follows that to end war or bring peace to earth the argument had to be resolved and the polarities reconciled. An understanding of change which accorded with the instinctively known absolutes had to be reached, which now is at last achieved. Above all we have had to find the true reason why we have been competitive and aggressive in order to explain our upset and establish our dignity. Finding this reason marks the arrival of

> **War is over,**
> **peace has arrived**

peace to earth. The realist and idealist, the right wing and left wing, the young and the old and, above all, the ambivalence within ourselves, all can now be reconciled. The so-called struggle between 'good' and 'evil' is resolved. Our divisiveness is now explained. Humanity has won its fight. The human condition is resolved and humanity is at last Free.

PART THREE
Conclusion

The Role of Science and the Mechanistic/Objective Approach

NOW THAT THE CONFLICT between 'good' and 'evil' is resolved it is possible to clearly explain the different roles mechanists and holists played as humanity developed through its adolescence.

There have been two ways of approaching knowledge, objectively and subjectively. Objectivity employed experimentation to establish what was correct while subjectivity or introspection employed our soul's conscience to know what was right.

Through our conscience we have always known the absolute truth or ideal of integrative meaning. The problem has been that without explanation/justification/reason for our divisiveness — for our anger, aggression, competitiveness and selfishness, our apparent inhumanity — this integrative/selfless ideal was impossible to live with, which left us no choice but to evade/repress/block-out any awareness of its existence. We had to find understanding of why we were divisive before we could confront integrative meaning/God. We had to defend ourselves

172

before we could confront God. We have been evasive or God-fearing.

To understand/explain ourselves we would need to understand our world that produced us. For instance, we would have to discover how the genetic and mind-based learning systems worked. The extremely difficult task was to do all this while evading integrative meaning because all understandings related to and led to integrative meaning since all development was a product of integrativeness.

Of the two ways of finding knowledge only the objective/ mechanistic approach allowed us to be evasive. Subjectivity/ introspection involved using the guidance of our conscience's orientation to integrativeness which meant living with integrative meaning which we could not do. Our instinctive, genetically based conscience could recognise what was and was not integrative or Godly but relying on it to do so entailed confronting integrative meaning. Introspection or subjectivity used the conscience to sieve out any understandings which were inconsistent with integrativeness. It worked downwards from the absolute truth to the particular. Living with the absolute truth of integrativeness, as subjective inquiry did, was hurtful and thus dangerous. So we had to be mechanistic rather than holistic. (As has been mentioned before 'holism' is defined in the *Concise Oxford Dictionary* as 'tendency in nature to form wholes . . .'. So holism involved recognising integrative order or wholeness, which we could not do.) It was only objectivity that allowed us to work from the particular, such as the mechanisms of change in our world, up towards an understanding of the absolute truths. Mechanistic inquiry was reductionist. It started from or reduced everything to the particular and worked up from there, while introspection worked down from the whole to the particular.

For two million years all humans have been alienated, to varying degrees. Our alienation was established in two stages. In the first we were born into a world instinctively expecting to encounter the sort of treatment and environment to which our original instinctive self or soul had become accustomed during its ten million years of indoctrination in love. But ever since humans first became upset two million years ago as a consequence of setting out in search of understanding it has not been possible to have our expectations met. To varying degrees infants have not received as much love as their souls expected and as children they have not been reinforced as much as their souls expected. In general we have not been able to grow up surrounded by non-egotistical, non-angry, non-arguing, non-alienated and non-superficial people in a natural environment in line with that expected by our souls. Reaching adulthood after such a childhood we were faced with bringing a new generation into the world without being able to defend/explain and thus give these new children any more understanding than we had received ourselves to cope with the apparent 'mistreatments' of our soul. And so on, generation after generation. As children the only device we could employ to cope was block-out or repression of the upsets, the hurt, the 'mistreatments' of our soul. We could not hopelessly keep on trying to confront the pain — the 'wrongness' of what had happened. This blocking out of, estrangement from, our true situation and thus true self — and the consequent partitioning of all the truths that our soul could guide our thinking towards — was the first stage in the establishment of our alienation.

Not being able to defend our alienated state, we have not been able to admit too clearly what alienation was. The immensely differing degrees of upset we each experienced as infants and children and the extreme sensitivity and vulnerability of our souls to 'mistreatment' when we were so young are subjects we have had to evade. For example, the more exhausted adults became searching for understanding the more they had to live with a conscience unfairly criticising their behaviour the more they needed to be told they were 'good' and not 'bad' the more unjust any reinforcement and admiration of others, especially innocents, seemed. But children came into the world expecting to find adults capable of admiring their efforts. Since parents were unable to explain their embattled state children could only interpret this lack of reinforcement as implying they were not deserving of admiration which made them die inside themselves. If you 'walk into a room full of people who are starving [in this case bereft of self-esteem] you can't expect to be fed' but children weren't able to be told 'the room was full of starving people'. Quite the reverse happened. Adults pretended there was nothing wrong. They laughed, joked and played and said 'what a wonderful day it is'! The truth is our adult world was a mad place, a nightmare to a child. Today's children are born somewhat adapted to (selected to cope with) this mad world of adults but this only hides the extent of the madness of it and thus a child's vulnerability to it.

Now that we can defend our battle-weary state it is possible to look at the extreme vulnerability of our soul and the immense differences between people according to their degree of alienation. When we do, we will be shocked at the depth of the upsets within us and at the immense difference in alienation between people. While we haven't been able to admit it until now, alienation was the real difference that existed between people, was the reason for our different personalities as it was the real difference between races and cultures. This threat of exposure to the differing degrees of alienation can make the more alienated among us afraid of exposure/revelation/'judgement' day unless there is sufficient counselling available to explain and bring us to an understanding of our misunderstandings/insecurities/worries.

The second stage of our alienation was a product of the first. Unable to defend/explain the upset alienated state and its accompanying 'corrupted/distorted' behaviour brought about by our imperfect childhoods, we again had to resort to blocking out in order to cope when we became aware adolescents. As adolescents and adults we had to block-out or evade the criticism such partial truths as integrative meaning made of us. We each blocked-out or hid from these hurtful partial truths to different degrees, according to how alienated we had become during our early upbringing. We lived at a distance from them commensurate with our particular degree of alienation, using differing degrees of false argument to defend our 'corrupted', non-'ideal' state. The greater truth actually was that we were always committed to integration even though we appeared to be divisive, but until we understood the truth and found the explanation for our divisiveness we had no choice but to be evasive, which added to the alienation we acquired in our infancy and childhood.

So the amount of truth one human could confront was different to the amount another could confront. This made collective introspection particularly dangerous since it would involve confronting many people with more truth than they could cope with. Therefore, as a community or collectively, in order to be safe and not expose anyone beyond their capacity for exposure, subjective/introspective inquiry had to be avoided rather than cultivated. For this reason, we were unable to establish the sorts of centres, such as universities, for introspection that we established for objective inquiry. It also meant that our communal vehicle for inquiry, science, had to be mechanistic rather than holistic/vitalistic. As a community we could not afford to recognise the vital force implicit in the tendency in nature to form ever larger and more stable wholes.

The great benefit of evasive mechanistic science was that it allowed us to work towards finding our freedom (from criticism) while avoiding criticism or hurtful exposure. It was the one way we could live while we carried out the search that would bring us our freedom. The alternative of subjective inquiry was too dangerous to pursue collectively. So science got on with the job of investigating the particulars or mechanisms of existence while avoiding the whole or overview. It was the only safe, nonthreatening approach to understanding available to us.

Ultimately, by learning to understand ourselves and our world, we would learn why we had been divisive and, with this understanding, we could at last live with and acknowledge the fact of integrativeness — we could at last confront God. The great task of investigating the mechanism of existence was the role of science. Its responsibility was to be evasive.

The Role of the Holistic/Subjective Approach

While science was mechanistic or objective in its approach this does not mean that science did not employ introspection. It did but it could not admit to this without bringing unjust criticism upon those who could not confront integrativeness, be integrative and think integratively. In truth, as Einstein pointed out in his quotes given earlier, effective science depended on an equal blend of objectivity or experimentation with soul-guided introspection. To use Professor Birch's comment, also mentioned earlier, effective science depended on a strong conscience.

The main talent required for the research or experimentation that objective inquiry depended upon was the ability to reason or think — mental cleverness or speed at mental information processing or I.Q; the talent subjective inquiry or introspection depended upon was soundness of self or lack of alienation or access to our soul's guidance for thinking. People who were exceptionally good at introspection — who were innocent of hurt to their souls — who were exceptionally sound thinkers — were what we once called prophets. So in truth the ideal or most effective scientist was someone who was in part a prophet. However, we could not acknowledge this or take steps to select scientists for this quality. We could not afford to recognise or acknowledge soundness (because of the criticism implicit in being 'unsound') until we could defend our battle-weary/exhausted state. We could not establish soundness tests to accompany our I.Q. tests to judge who should enter universities. As has been explained in Part 2 of this book, had alienation tests been carried out along with I.Q. tests we would have discovered that the more intelligent were in general the more alienated; therefore, to find the ideal blend of soundness and cleverness, we should have chosen those with average I.Qs. The reality however was that we were unable to recognise soundness and by stressing only cleverness we became extremely unbalanced, as will shortly be explained. (It should be acknowledged that some scientific subjects were more remote from the human condition than others making them less threatening and thus less in need of soundness to investigate them. Subjects such as mathematics and astronomy were relatively 'safe' non-threatening subjects which did not suffer greatly from being investigated by alienation while subjects that brought the human condition into focus, such as biology, required the soundness of people like Darwin to investigate them effectively.)

By the way it was evasive to claim that the reason we didn't assess soundness is that we didn't know who was sound and who wasn't. We repressed soundness and in the extreme crucified it, which we could not have done if we were not able to recognise it. In fact it would have been easier to establish tests that measured alienation than it was to establish tests that measured I.Q. Teenagers at school have a number system for 'sex' which goes: 1 is holding hands, 2 is kissing, 3 is putting your hand under the blouse, 4 is putting you hand up the dress, etc. These steps in perversion development go on and on and provide an accurate measure of our degree of upset/anger because our level of upset/anger and with it alienation is reflected in our degree of sexual perversion. Sir Laurens van der Post, in his book, *The Heart of the Hunter* (1961), mentions that 'Not by the men, but by the women who flock to him and their obedience, shall you first know the true prophet'. Men normally 'used' women and when they did not, and it was apparent in their eyes whether they did or not, women recognised it. Women were the first to recognise and acknowledge an innocent — a prophet.

In earlier, more naive times, human societies often did recognise and make a place for people who were exceptionally good at introspection, able to confront the truth and reveal it to others without evasion — who were exceptionally free of the need to be evasive — who were necessarily exceptionally innocent of encounter with the compromise of the human condition and who were thus unhurt, uncorrupted, sound and not upset. The Hebrews of biblical times fostered such people — they were the prophets of the Old and New Testaments. More recently we have become too sophisticated or aware of the dangers acknowledgement of prophets and their sound thinking brought upon us to openly recognise and collect them. Their great innocence of hurt, their soundness, unevasiveness and openness only served to remind us of our non-innocence, criticising us and increasing our hurt.

We certainly needed innocence/soundness/introspective-guidance/subjective-capability/conscience but it also served to criticise us. We had to find the full truth that defended us before we could even acknowledge the existence of soundness and innocence let alone openly select for it and cultivate it as we could the cleverness required for objective inquiry.

The Limitations of Having to be Mechanistic

Progressing evasively, mechanistically, while it was the only approach we could take, nevertheless had severe limitations and consequences which now have to be confronted.

By stressing only objectivity/mechanism/reductionism/cleverness/evasion and evading subjectivity/holism/expansionism/soundness/confrontation, science was bound to become unbalanced. The result was that it ended up extremely insecure with an accumulated mountain of evaded truths.

Starting out God-fearing science would end up God-terrified. It would slam the door on anyone trying to acknowledge integrative meaning. When this author naively went to London in 1983 to personally submit a summary of the ideas in this book to *Nature* magazine (one of, if not <u>the</u> leading scientific journal in the world) for publication the door to their offices <u>was</u> all but slammed shut against him when he mentioned integrative meaning.

Its insecurity aside, the main problem science created for itself with its evasive approach was: how did it hope ever to dismantle the 'mountain of evasion' it had accumulated; how was the full liberating truth to be extracted from all these evasively presented insights. Science did a magnificent job of finding the mechanisms, all the pieces of the jigsaw of explanation, but presented them upside or picture-side down because it could not look at the picture while it was incomplete. But this made it impossible for science to gain an overview, to assemble all the pieces of the jigsaw. It was not possible for humanity to find its freedom this way. It was unable to be unevasive through its science.

The full truth about ourselves could only be found by confronting/living with/accepting the many hurtful partial truths that most of us were evading. That was the final paradox. For example, the process of love-indoctrination could not be recognised if love or selflessness was not seen as meaningful. The whole jigsaw of explanation would not come together unless integrative meaning was accepted. The full truth could only be built from the many hurtful partial truths that science was evading!

The Need for a Prophet to Liberate Humanity

To solve this problem what was required was for someone to come along who was exceptionally unevasive, someone with an exceptionally strong and thus good conscience for sieving the mind's thoughts, who would defy and thus not adopt all our evasions. This person, by retaining only the truthful components of science's hard-won insights, would assemble the full unevasive truth for us. To defy all our evasions this person would need an exceptionally unrepressed conscience. He would need to be exceptionally sound. The only people who are not at all evasive of the truth, who do not fear integrative meaning/God, who 'delight in the fear of the Lord' (Isaiah 11:13) are our prophets, those among us who are exceptionally free of 'mistreatment' or hurt (encounter with the compromise and upset to our soul associated with the human condition) who have no cause to evade/repress/alienate their soul and its conscience. The difficulty with this is that humanity has not been able to cultivate soundness/introspection/subjectivity/innocence. We have not cultivated prophets at all, let alone an exceptional one. Quite the reverse, we have repressed them. As mentioned, this is the final manifestation of the basic paradox of the human condition: we do not want the truth but then we do want it! We had to be evasive only to end up wanting someone who was not evasive! We have been destroying what we wanted! We have been unable to collectively cultivate innocence and yet only innocence could confront the truth — could live with the hurtful partial truths that science avoided — could afford to be unevasive — could look at the pieces of the jigsaw right-side up and thus assemble the full picture — could liberate us.

The result of our evasive, objective approach has been that we have been able to do no more than have 'hope and faith' that, when we had evasively revealed enough of the mechanisms involved, a person capable of exceptional subjective or unevasive inquiry would just appear. We had to just hope that somewhere in some isolated/sheltered/protected corner of the earth a man would grow up with his soul intact and liberate us. So while the objective evasive approach had to be employed throughout our journey to find understanding in the end we required exceptional subjectivity or introspection or unevasive talent to liberate us.

He would have to be a 'he' and not a 'she' because, as has been explained before, while women were capable of being exceptionally honest, not being responsible for overthrowing the threat of ignorance they were not in a position to reconcile the upset on earth because at its base the upset was caused by the threat of ignorance. However women had a vital role to play. To produce such exceptional innocence required the occurrence of a woman capable of giving her son only pure love. Since women were made aware of the upset on earth through 'sex', what was needed to produce a mother capable of giving a son only pure love was a woman who was not aware of this upset, who had not been exposed to 'sex', to 'fucking', to the destruction of innocence in women. The image is of a virgin, hence the image of the virgin mother Mary of the exceptionally innocent Christ child. While an exceptional prophet/innocent had to be a male an exceptional mother was required to produce him. The primary role of women was nurturing while the primary role of men was either fighting ignorance (if they were exhausted) or (if they were innocent), fighting/defying evasion. This was a fortunate arrangement since fighting and loving are opposite qualities and thus incompatible in the one person.

The Problems Confronting a Liberating Prophet

The task confronting the liberating prophet was formidable because he had to overcome all our evasions not just some of them. Some truth was dangerous while the full truth is not. The full truth which explains why we humans have been upset and divisive is compassionate. It does not criticise us. However, as is revealed in this book, that truth can only be arrived at via all the partial truths such as integrative meaning which on their own criticised us. Similarly, viewed separately the concepts of love-indoctrination and development criticised us. But when these ideas are presented as part of the full story as they are in this book they do not criticise us. It was a case of having to deliver all the truth or no truth — no critical truths anyway — if we were to be liberated from criticism rather than merely adding to the criticism with which we had to live. We needed from a prophet composure not exposure. We needed our sense of guilt alleviated not increased. We have lived in fear of dangerous/hurtful prophets, prophets who have left us exposed to criticism. We wanted a prophet who would lead us all the way through the minefield of hurtful partial truths, through the potentially devastating criticism, to safety on the other side, where the full truth and compassionate explanation for our divisiveness would be found. Humanity has been milling at the edge of that minefield in ever increasing numbers but could not venture out into it. Using another analogy, only a David (an exceptional innocent) could go out from the besieged ranks of humanity's army and slay the giant Goliath (ignorance).

The formidable task facing the prophet was to defy all our evasions and hold onto all the critical partial truths in the face of the determined and normally overwhelming efforts of his fellow humans to keep them repressed. Only by doing this could he find the full liberating truth. To be able to do this, to stand by what his conscience said he should believe instead of what the rest of us told him to believe, he needed an exceptionally strong conscience. He needed to be exceptionally innocent of hurt/'mistreatment' and thus free of alienation and thus secure or uncorrupted in soul and conscience. He had to be so secure he would not be swayed by our evasions/falsehoods/distortions/confusions/upsets/denials/superficialities.

In his younger years especially the world of a prophet was one of fight — of defiance — of hanging onto what his soul was telling him was right in spite of what the world around him was telling him to believe. A prophet was not like a saint. The *Encyclopedic World Dictionary* says a saint is 'one of exceptional holiness of life' and a prophet is 'one who speaks for God'. A saint lived a holy life while a prophet spoke of holiness. A saint was passive while a prophet was active. If Saint Francis of Assisi, loving the animals as much as he did, was strong/secure enough he would have taken that love into battle and tried to solve the 'wrong' on earth. The problem that had to be solved if animals were to be saved was our human upset. Necessarily saints were people who had lost their innocence but been 'born again' or gone back to purity of life in an exceptional way. On the other hand prophets were necessarily people who had not had their innocence spent or destroyed. To speak for soundness, to get the truth up, the soundness had to be very strong and clear in you. In fact it had to be you and not something you had returned to. Leonard Cohen in his book, *Beautiful Losers* (1966), made the distinction when he said:

> 'A saint is someone who has achieved a remote human possibility. It is impossible to say what that possibility is. I think it has something to do with the energy of love. Contact with this energy results in the exercise of a kind of balance in the chaos of existence. A saint does not dissolve the chaos; if he did the world would have changed long ago. I do not think that a saint dissolves the chaos even for himself, for there is something arrogant and warlike in the notion of a man setting the universe in order.'

Another difficulty confronting a liberating prophet was that he had to work alone because we were unable to cultivate/institutionalise the art of introspection or even recognise its unevasive thoughts. Further still, he could not reveal and find recognition and support for his work until he had found the full compassionate truth and even then, in our shock at being confronted with the truth, he would have to expect to be assailed rather than supported for the work he had done.

As well, throughout his battle to defy us he would find no appreciation for the subjective way he thought, being the complete opposite to the normal mechanistic approach. The mechanistic approach was detail-particular and whole-view-uncaring or unconcerned while the subjective approach was whole-view-caring and detail-uncaring. For example, a holist could not very well study mathematics or chemistry or learn to spell or be concerned with grammar while the questions of why people were unhappy and even starving were not being addressed. He could not let go his belief in integrative meaning. While this author did finally gain a science degree in zoology after attending two universities it took me five years to complete the three-year course. In that five years I had to repeat the compulsory first year chemistry and mathematics three times before I passed them and now, forty-one years old, I still can't spell or use grammar correctly. For this book a great deal of editing (400 hours) was required to make (I hope) my holistic way of thinking meet our mechanistic expectations. My school, Geelong Grammar in Victoria, is world famous for the special effort it makes to foster innocence/soundness. To quote from an article in the school's magazine titled *What we profess and practice*: 'Primal innocence, like primal Eden, is destroyed: yet both can be restored; the Divine Image lives on, the burden and the glory of mankind, and true education consists in its recognition and its restoration . . .'[1] Geelong Grammar held no entrance exams and was one of the first schools to go outward [nature] bound, do away with uniforms, play down competition in work and play, scale down disciplining military cadets, and go co-educational. Yet even this school told my parents that I 'would find a science course at university too difficult'[2] and encouraged me to do manual subjects — tried to exclude me from humanity's search for understanding.

We can get a glimpse here of just how much humanity had overshot the mark in terms of stressing cleverness and in the process excluding soundness. Through the support of my parents I was able to qualify for university by studying and passing the State Leaving Certificate exams by correspondence from

---

[1] *The Corian*, April 1982.
[2] From the author's last school report card in 1963.

home on the family sheep station/farm. I was always in the lowest classes through school and at university never gained better than a pass in an exam. Proof that it was the false mechanistic evasiveness of academia that was the problem was that on the one occasion I was allowed to think holistically, in the biology exam for the Leaving Certificate, I gained first class honours and came fiftieth in the state of New South Wales. In the exam I was allowed to write an essay of my choosing and I wrote about 'Why don't some ants become lazy and live off the rest'. It is amongst ants that co-operation/integration is most apparent or least deniable in our world. A prerequisite for success in academia as it has existed was alienation. Tragically, soundness was not able to be considered. I still have nightmares about exams.

While we have had to be evasive the holistic truth, which can now be admitted, is that the incredible importance placed on the fabled 'three Rs' (reading, writing and arithmetic) of the mechanistic education system <u>was</u> absolute bullshit (evasion) compared to the need to teach love, enthusiasm and happiness. In truth our schooling was all about introducing children to death, not life. While death of soul and spirit was the reality of life during humanity's adolescence and this had to be prepared for 'true education consisted of the recognition and restoration of primal innocence'.

The holist's ('wholist' seems more appropriate) world was the complete opposite of the mechanist's world. You assessed the quality of a holist's work according to how accurately he or she had understood the whole view. You assessed the work of a mechanist by judging how accurate the details were. Mechanists could have the details perfect but the whole view in complete disarray and be happy. Holists could have the whole view perfect but the details in a complete mess and be happy. The priorities for a holist were the opposite of those for a mechanist. The more exhausted or battle-weary we became the more the boundaries of our world were reduced. In the end our world often became so limited we were only capable of keeping our car nicely polished or our room perfectly organised or our dress immaculate; everything beyond these limits had to be ignored and left in chaos. On the other hand, if we were completely unembattled the boundaries of our concerns would not have been reduced at all in which case we would not have been able to sleep at night knowing there were people in Ethiopia who were starving and the last thing we would have been concerned about was whether our hair was done and 'how we looked'. It should be stressed that lack of pride in self did not necessarily indicate a holist and similarly the desire for perfection of detail did not necessarily imply a mechanist. We are talking generalities here. Someone who makes furniture (as this author does for a living) could try to design furniture in a holistic way, free of embellishment/extravagance/self-glorification/ego and concentrate on trying to find the simplest and most natural design possible. He could pursue confrontational, thoughtful, sound solutions rather than escapist, thoughtless, therapy solutions. When looking at the furniture that resulted, mechanists would evasively not see the simplicity, ingenuity and soundness in it and instead see it as boringly plain. Further, mechanists would prefer furniture that was superficially perfect to that which was genuinely perfect. They would prefer furniture that had an immaculate finish and was built with junk materials rather than furniture that was profound and had little care taken with the finish. Mechanists and holists had completely different concerns. Recalling Sir Laurens van der Post's quote on the sensitivity of the Bushmen of the Kalahari, a holist (in this case the Bushmen) could 'know what it felt like to be a Baobab tree' but not know the botanical name of it while a mechanist could know the botanical name of every tree on earth but have no feeling for trees at all. In studying the relatively innocent Bushmen, van der Post himself encountered the extreme superficiality of the evasive mechanistic world. In his book, *The Lost World of the Kalahari* (1958), he says: 'I found men willing enough to come with me to measure his head, or his behind, or his sexual organs, or his teeth. But when I pleaded with the head of a university in my own country to send a qualified young man to live with the Bushman for two or three years, to learn about him and his ancient way he exclaimed, surprised: "But what would be the use of that?" ' A holist dietician could well enthuse about people going onto a vegetarian diet (to say that we should be vegetarian is holistic and not mechanistic/ evasive because it confronts us with our post-innocent angry/up-set lifestyle where we changed from being vegetarian to killing animals and eating meat) without being concerned that he or she did not know what vitamin E or cholesterol was. On the other hand a mechanist dietician would talk endlessly about the existence in the intestine of cholesterol-colloid-mouse-umbrellas or some such extraordinary substance without ever looking at the real dietary concern for humans of getting back to a natural diet. As well, the mechanist dietician would likely heap scorn on the holist dietician for not knowing what vitamin E and cholesterol were and the holistic dietician, not being aware of the way mechanists think, would have no idea why he or she was being criticised over matters that are to his or her way of think-ing not the priority concern at all. It has been a mad world to live in. Thank heavens the madness can end. Holists and mechanists were worlds apart and neither understood or appreciated the other's way of thinking at all. Christ's description of mechanists as 'blind guides' who 'strained out gnats but swallowed camels' (Math 23:24) was accurate. Again it has to be remembered that until we could defend ourselves in the presence of the truth we had no choice but to evade/escape the truth — to be mechanistic. To be evasive was the correct procedure. Holists were dangerous because they unfairly criticised/exposed us.

As mechanists we had extremely superficial measures for what represented quality, we didn't think straight and we were non-lateral or unimaginative and, above all, insecure in our thinking. Living with the whole view was an entirely different approach to evading the whole view. Not being aware of the art of living with the whole view we measured the quality of holis-tic/subjective thinking in our extremely restricted terms and tried to force it to adopt our extremely limited way of thinking.

As has been explained, the world of innocence was a totally uncared for and uncatered for world.

Finally, to return to the task confronting a liberating prophet. When he found the full truth the prophet had to try to present it to us even though we did not want to hear it. We did not want to hear it because even though it was compassionate it still involved confronting many truths about our world and ourselves that we were trained to repress/evade/deny and as well it meant a challenge and complete readjustment to our way of thinking which are in themselves traumatic events. Revelation day is also 'judgement' day, exposure day, confrontation day and readjustment day!

While the task of a liberating prophet was formidable, to help him he had the power of love — of access to the beauty, truth and happiness of our soul's world. Also, he could share his thoughts with and derive comfort from integrativeness/God. In truth the strength that could be derived from the wonder and happiness of the true world that our soul had experienced and knew all about was a force so strong that our exhausted, battle-wearied world could not overcome it no matter how incredibly great the fury of a mind unjustly criticised. In our upset we could attack the earth and even have destroyed it but still love would have survived and remained untouched. What is everlasting or immortal and profound or unmovable in our world is integrativeness or love.

Of course, having said this, we have to remember that the fury of a mind unjustly criticised was due to the fact that it was unjustly criticised, that it was fighting for the permanent establishment of love on earth, that its work was for the benefit of integrativeness (even though paradoxically it did not appear that way) and was thus of everlasting and profound significance. The spirit of man, which has been passed on from generation to generation for two million years, was committed to establishing recognition of its 'goodness'/love/beauty/worthwhileness/Godliness — to its immortal and profound significance.

The Restoration of Innocence

It is now an opportune point to put an end to our evasion and exclusion of innocence and their world.

Although mechanists have until now been misunderstood we are nevertheless familiar with them and their world. Holists on the other hand have had their existence and world evaded or, if we couldn't evade them then we made them so exalted — surrounded them with so much religiosity — that we distanced ourselves from them, alienated them from our presence anyway. Until now the world of innocence has criticised our lack of innocence and so we have preferred not to think about it. Now that our legitimate exhaustions are defended we can and must consider and foster the innocence that resides both in ourselves and in others. It is the way out for humanity.

The truth is, exceptional innocents or prophets were neither superior nor supernatural beings but people like ourselves with their own particular encounter with the human condition to contend with. Prophets and innocents in general can now be recognised, understood, humanised and brought into our presence. As we have to rehabilitate our soul so we have also to rehabilitate its manifestation which are those among us who are innocent.

To take the first step towards demystifying and restoring innocents, it is necessary to explain the life of exceptional innocents or prophets.

In the first place innocents did not knowingly set out to reveal the truths in our evasions. Someone innocent of the battle associated with the human condition had no need to be evasive and so, not experiencing the need for evasion or knowing the reason for it, naturally resisted adopting evasion. The effect was that the potential prophet listened to what his conscience said and not to what those around him said. Doing this was harder than it may seem because, being innocent, he would not be aware initially that those around him were being evasive. He would trust us, presume us sound, since he knew of nothing else, and we, having to evade the fact that we were exhausted from the battle, could not tell him we were no longer sound. We could not tell him that we were 'false' and that he should not trust us.

When we were exhausted we evaded, blocked-out, the fact. Alienation was a very real thing. It was unaware of itself. If we were aware of our state we would not be alienated, we would not have blocked out the fact of our lack of innocence. Being alienated meant that we were not inclined to be aware of innocence and its world. In these circumstances we were often unaware that we needed to defend/explain or even admit our exhausted state. Evading the fact that there was another 'ideal', integrative state we all too often saw our embattled state as being natural for humans and therefore self-evident to others. On the few occasions when we did realise our situation (such as when we were unavoidably confronted with innocence whether in the form of an innocent person, an innocent act or an innocent/unevasive thought) we were unable to defend it anyway and so had to return to being evasive and blocking out the truth of our condition again. It was our inability to explain and thus understand ourselves and the 'mistreatment' that had been inflicted on us that caused us to block out in the first place.

However, our alienated state and the reasons for it were not 'self evident' or apparent to an innocent. Not having experienced the battle and the need for evasion and unable to be told about it, the innocent were left with no way of appreciating the world of the battle-fatigued. In fact it was almost impossible for an innocent not to succumb to the reality of this world. (Of course now that we can defend/explain ourselves the alienated can be rehabilitated and innocents told about and thus protected from alienation.)

The great mystery innocents faced was why people were behaving so strangely, so unnaturally, in terms of what their instinctive self or soul (which we all had but which was unrepressed/uncorrupted in an innocent) expected. The more innocent (the less repressed the conscience), the greater this mystery. Look at prehistoric cave paintings. While adolescent prehistoric man could draw the animals around him with total empathy, when it came to portraying himself, especially his face, he had no empathy and resorted to clumsy stick drawings. (To quote the book, *Gardners Art Through The Ages*, fifth edition, 1970, '. . . animals are rendered with all that skilled attention to animal detail we are accustomed to in cave art . . . But . . . man is rendered with the crude and clumsy touch of the un-skilled . . .' Why?) Exhausted/embattled/upset man was some-one our innocent instinctive self had no understanding of. The shyness we have always associated with innocence was due to their lack of familiarity with what was going on in the embattled world. So great a mystery was this world to innocents that they spent the first half of their lives trying to make the exhausted admit to their strange behaviour, to make them stop 'lying' (this being the only interpretation innocence could put on our eva-sions), to make them 'own up' or repent or be honest. Christ, an exceptional innocent, began his search for understanding of the world with this preoccupation. As it says in the Bible, 'Jesus began to preach "Repent" ' (Math 4:17). While many of the battle-wearied did acknowledge their state and decide to live through Christ, the real problem,the reasons humans needed to be evasive, remained unsolved. Christ wanted us to stop being evasive but couldn't force us, which was why he became upset and 'denounced' those who 'did not repent' (Math 11:20). To young innocents evasion of the truth was unnecessary and corrupting. They could walk around freely in the realm of the truth. They did not fear God and if the exhausted could not tell of their fear of God the innocent could only interpret their fear as denial of the existence of God. It has been a horrible and tragic situation for all concerned but now at last it is ended.

So all an innocent had to help him hold on to his ideals and resist succumbing to our evasions was his original integratively orientated instinctive self, or conscience. Obviously if he were to hold onto his ideals — such as a belief in integrative meaning — against the sea of evasion around him, he would need an

exceptionally strong conscience. As mentioned, to be this innocent — to have this strong a conscience — he would need to
have encountered only pure love from his mother in his infancy.
He would have to have had a mother who had not been made
aware of the upset on earth, who was, symbolically speaking, 'a
virgin' (Math 1:24). In his childhood he would have to have
grown up relatively isolated from our battling ways and in an environment that was familiar to our soul which is an environment
of nature. The background necessary to produce such exceptional innocence and how that innocent behaved subsequently is
clearly described in the Bible. The 'child' will 'live in the
desert' . . . 'with the wild animals' . . . 'eating curds and honey'
where he 'will grow and become strong in spirit', grow up with a
mind guided by a strong conscience, strong enough 'to reject the
wrong and choose the right' (Luke 1:80, Mark 1:13 and Isaiah
7:15). Only a very strong/clear conscience would refuse to
waver/succumb/be defeated by the evasive world's coercion to
compromise with the absolute truths.

The innocent's experience of the paradox of the human condition was that, while instinctively he wanted to trust and support the wishes of his fellow man his instincts/conscience were
also telling him that what we were saying, our divisive biological
theories,for instance, was wrong. As a young man, an exceptional innocent had only his conscience to stand against the
world of evasion. Otherwise unarmed, he could do nothing but
stand there, face black with defiance, trying desperately to hang
on to what his conscience was saying in the face of the almost
overwhelming coercion from the rest of us to compromise, to be
evasive. This phenomenon is described in the Bible where it says
that as a young man a prophet would be 'consumed by zeal'
(John 2:17 and Psalm 69:9). Initially zeal or enthusiasm for the
'true'/ideal world was all a prophet had to hold on to but gradually he would learn that what he thought was profound and what
the world of reality was proposing was often superficial/evasive/
false. He learnt that 'If I do judge, my decisions are right' (John
8:16). It was only natural that the first to bear witness to his

soundness, to experience or discover it, would be himself. However, only when he eventually worked out why the rest of us were being evasive, only when he learnt 'what was in a man' (John 2:25), could he afford to let his zeal (his determination to defy our ways and beliefs) subside. He had to work out why we were being evasive without ever succumbing to our evasion. Alienation could not investigate alienation. For example, once the central truth of integrative meaning is evaded we have adopted a false premise from which to continue all thinking. From such a standpoint, sound thinking is impossible. Only innocence could reconcile the world. A prophet had to 'overcome the world' (John 16:33) and its 'lying' (evasion), not succumb to it, if he were to find a truthful reconciliation/understanding of our upset condition. Of course, with the truthful defence for our embattled state found, no one need be evasive any longer. This brings about the end of upset on earth.

Incidentally there have been three types of prophets and these can now be identified. There have been false prophets — those who had been 'born again' back into the world of our soul from a post-battle state who were thus imitating soundness, pretending to be able to 'see' and guide us when they could not. (The position of false prophets should be contrasted with that of scientists who were not dishonest about their soundness, they have not claimed to be able to 'see' and 'guide' us, they were mechanistic/objective, relying on empirical evidence to establish what is truthful, and made no claim to be holistic/subjective/sound.)

Then there were the real but hurtful and thus dangerous prophets. These were the genuinely innocent who had not departed very far from their soul and who were thus still close to its world but were not sufficiently close — sufficiently sound — to reach all the way to a full unevasive understanding of the human condition. It was through hurtful or dangerous prophets, from those of biblical days and earlier right down to the modern prophets like Eugene Marais and Sir Laurens van der Post, with their revelations of many hurtful or dangerous partial truths, that the non-evasive truth was able to 'forcefully advance' [Math 11:12]. Their work (plus, it should be added, some of the less weird, less superstitious, less mystic efforts of the false prophets to be unevasive) helped open up a path for subsequent prophets. If a prophet had to be entirely alone with his unevasive thoughts it would become too difficult to hang onto them in the face of the opposition from the rest of the world. He was being led by his conscience into a realm the real world denied existed. The knowledge that someone somewhere had had the same thoughts was enough to let him know that he wasn't mad to think these thoughts and this enabled him to hang onto what he was thinking and continue thinking.

The problem for a prophet was not in finding ideas so much as finding the strength to dare to think exactly what his conscience wanted him to think. (As an example of this when the thought first occurred to me that the highly revered and worshipped 'person' we call God was none other than the theme in nature of integrativeness I gave myself quite a shock and thought I should reject the idea — a reaction I had to ride out before the idea could gain a solid footing in my mind. Some ideas were so difficult to 'let through' they had to be written down and then typed up and, as it were, forced into the open and made to stand up straight. They had to be 'dragged out kicking'. Above all finding the explanations presented in this book was an exercise in learning to stand by exactly what my conscience wanted to say, was learning to trust my conscience and not those around me.)

The third variety of prophets were compassionate prophets. These were a variety of real prophets that could not only reveal repressed partial truths as dangerous prophets could but could reach all the way to the full compassionate truth that was built from repressed partial truths. Necessarily they were exceptionally innocent innocents who had not left the world of their/our soul. They had to have very strong consciences to enable them to not only defy our evasions but 'overcome' them by learning why we were being evasive. In the recorded history of humanity, there have been very few compassionate prophets. Christ was one although his explanation of the full truth had to be couched in metaphysical terms as we had not found the scientific knowledge that would make it possible for him to clearly explain his reconciliation of our condition. Only a first-principle-based defence of humanity could be understood and thus shared by us all. Metaphysical explanations such as 'God loves you' did not explain why God loved us — did not give us the ability to understand ourselves. For the unevasive full truth to be liberating it had to be in understandable first principle terms as it is in this book. This need for humanity to first find the knowledge that would make it possible for a compassionate prophet to liberate us will be explained in more detail a few paragraphs on.

It should be apparent from what has been said that in spite of his happy innocence the life of a prophet was as excruciating in its own way as was that of the embattled who had to put up with the horrible injustice of a constantly critical conscience. During humanity's adolescence, everyone — innocent, exhausted and all stages in between — suffered equally. As the rock singer Jim Morrison once said when describing life during what we now know of as humanity's adolescence, 'Nobody gets out of this world alive'! In truth everyone fought as hard as they could from their particular position in the battle to achieve liberation from ignorance. We were all equally good soldiers for humanity — all equally courageous.

Our Liberation achieved through the efforts of Everyone

This raises the main point to be made in this conclusion. While prophets were required to liberate humanity, the preparations which made liberation possible were a group effort, everyone throughout humanity's history has participated. The following analogy of building a staircase does not reconcile very well with the analogy already used of a mountain of evasion needing dismantling. However, with the mountain analogy it was the extent of our evasions which was being stressed. What is to be stressed here is the point that the physicist, Sir Isaac Newton, once made, that we could only get to see clearly by being hoisted up on the shoulders of others. Our liberation was made possible by the combined efforts of everyone. For two million years humanity has been sending wave after wave of its soldiers against the wall of ignorance and one day, upon the shoulders of all that effort, one of us finally had to scramble over the top of the wall and pull the bolt on the gate in that wall and let us through.

It <u>was</u> as if the whole of humanity worked tirelessly for two million years building a staircase up a gigantic wall (of ignorance) that barred the way. The bricks being used were our evasively presented understandings/insights. The completed staircase could only be ascended by an exceptional innocent who was unafraid of the view from the top of the wall, which was of integrative meaning. But this could not be done until the staircase was complete, until humanity as a whole had found all the mechanisms of existence. Only a first principle explanation could liberate us from our sense of guilt. In the time of Christ humanity had not yet found the insights that made liberation possible. For example, science had still to discover the mechanisms of genetics and the nerve-based learning system. In fact the discipline of science itself had yet to be formulated.

Christ was aware of humanity's need to find first the knowledge that would make liberation possible. He said: 'Although I have been speaking figuratively a time is coming when I will no longer use this kind of language but will tell you plainly about God . . .' and '. . . another Counsellor, the Spirit of truth, will be with you forever — will teach you all things and will remind you of everything I have said to you [the human intellect will carry out lots of research and find the everlasting or profound first principles or mechanisms behind the workings of our world. Then when this work is given to an exceptionally clear conscience a first principle defence of humans will be found that can be taught/explained/communicated to and thus shared or understood by everyone. When this happens we will be able to understand plainly what he, Christ, was only able to explain in metaphysical terms; we will see that our first principle explanation says the same things as Christ was saying.]' (John 16:25 and 14:16,17,26). Christ eventually had realised the limitations imposed on him by the times in which he lived. He realised that the most a compassionate prophet could achieve prior to humanity as a whole finding the mechanisms that would make liberation possible was to provide a place of soundness for less sound humans to retire to and recuperate after they became exhausted from 'building the staircase'. He learnt that while he could establish or found a religion he could not eliminate evasion, he could not stop us being evasive. The Jews acknowledge this. They do not consider Christ the liberating prophet, the so-called messiah, because they recognise that while he was able to temporarily 'deliver us from evil' — to give us 'salvation from sin' — he could not eliminate evil (upset) from earth. Humanity had to struggle on and find more knowledge before liberation would be possible. The staircase up the wall of ignorance had still to be completed.

In many ways prophets only got in the way while this work was going on because they depressed us by confronting us with truths we had no option but to evade. If exceptional innocence could have liberated us without knowledge having to first be discovered we could have liberated ourselves two million years ago because we were all exceptionally innocent then. Our exhaustions, which made the occurrence of innocence rare, came about because we had first to find knowledge and to do that we had to defy our conscience which upset/exhausted us. If anyone is tempted to think of exceptional innocents, their world and what they can do as being superior to our embattled world they have only to consider this: if exceptional innocents were left alone on earth, while they wouldn't have the battle-weary around them to corrupt them, they wouldn't have the knowledge the battle-weary found in the process of becoming battle-weary and they would eventually become as corrupted/battle-weary when they set out in search of knowledge, as they would have to do. The presence of 'corruption' was the presence of the search for knowledge. The exhausted were those who had already left innocence to do the work that had first to be carried out for humans to become free; they were those who had already gone to battle while the innocent were those who had yet to go. Innocence was in no way the hero of the battle. It had yet to even go to battle!

To use another analogy, that of a football match, to emphasise the point being made. Football teams have specialist players who kick the goal only after the whole team has worked the ball down the field to within kicking distance of the goal posts. Kept away from the main battle which has left the other players bruised, battered and tattered, this specialist player comes on the field in his clean togs with his hair in place (in some codes of football the specialist kicker isn't kept entirely out of the rest of the match but in others, such as American gridiron, he is) and takes the kick that scores the goal that wins the match. In the same way, while innocence came 'on the field' at the last moment to 'kick the goal' (liberate humanity) the victory clearly belongs to the players who did all the exhausting work that made 'the kick' possible. The reason we became exhausted is that we personally and those who produced us, namely our parents and their parents before them, etc., had been doing all the work — had been defying ignorance and searching for the understanding/knowledge that would make liberation possible. We had to 'lose ourselves to find ourselves'. The price of freedom was that humans be prepared to expend themselves in the battle, generation after generation, for two million years. Now that we are in the position of having to confront innocence and its world and accept its leadership out of our embattled/exhausted world it is very important that everyone remember this — that the glory of our success belongs to the battle-weary, those who did the work, those who expended themselves in order to achieve humanity's freedom, those who paid the price. We owe our success to them, they most deserve to be loved, they are our great heroes. This is the essential message of this book.

Finally, to return to the analogy of the wall of ignorance for a moment. When the exceptional innocent got over the wall he had to resist opening the gate until he had laid out the complete banquet — a feast of explanation. There had to be no gaps left in the explanation that might lead to misunderstanding, confusion and further upset. After thirteen years of introspection this has been done. The full unevasive story from atoms to humanity, without any gaps, has been thought through and is presented here. As has been stated so often, it was a case of all the truth or no truth — no hurtful partial truth, anyway. All the repressed truths had to be revealed together in one go to make the full truth. This inability to reveal the truth gradually and thus gently was a product of our necessarily evasive approach to the truth.

The full book that is to follow will not add new insights. Of necessity, they are all revealed here. What it does is expand on what has been sketched out here, elaborating in more detail on the scientific bases for the insights, and following them through to explain the many practical situations that confound us at every turn today. It spells out the ramifications, the full import, of this knowledge, which is not always immediately obvious. However, for the time being, there is ample explanation to adjust to in the condensed version. For example, having evaded the truth for so long just getting used to the idea of God being integrativeness and not some person in the clouds is shock enough. We have to become used to seeing clearly.

The Difficulty of Adjusting to the Compassionate but Unevasive Full Truth

While the evasive mechanistic approach was the only approach humanity could take it meant it would have to rely on 'hope and faith' that it could eventually liberate itself from the resulting mountain of evasion. Now that hope and faith have been fulfilled and the full liberating truth found one more serious problem remains as a hangover from this evasive mechanistic approach. It is how are we to accept and adjust to the truth now that it has arrived.

Given our evasive approach it was inevitable there would come a time when the mountain of accumulated evasions would have to be dismantled — our revelation or 'judgement' day. When we consider what this actually entails, we can see the magnitude of such a confrontation. Inevitably it must be traumatic. No matter how compassionate the revelations, the adjustment itself will be difficult. It will be a shock. We are accustomed to our old ways of coping, our old evasive defences and justifications for our conscious thinking self or ego, and can't easily accept their destruction and the adoption of a whole new way of justifying ourselves, of behaving. Men especially have had fragile embattled egos, as has been explained (which is the main reason it will be easier for women, who don't have embattled egos, to acknowledge the truth of, and act on the information in this book), which means men won't easily abandon the justifications they have been using to uphold their egos. Men quickly became so embattled, so 'punch drunk', that any step back or aside would seem to them like a retreat, like admitting defeat. The defiant expressions 'give me liberty or give me death' and 'winning is all that matters' are expressions of this madness. In order to go forward it is often necessary to go back, to change tack, to re-adjust, but the male ego could become so embattled men could refuse to change the defensive position they have adopted and become entrenched in.

We are all being asked to re-model our whole life with a new way of justifying it. As has been explained earlier, our personalities are the expressions of the way we have coped with life under the pressures of the human condition (of living with an unfairly critical conscience) and we can't be expected to change our personalities overnight. We would have to expect that many among us, especially the more ego-embattled and more exhausted with most to adjust to and be exposed will tend to resist such confrontation/exposure. As it says in the Bible, albeit in our old often critical metaphysical terms, about those who fear exposure 'Everyone who does evil hates the light' (John 3:20) and about those unable to re-adjust their thinking 'new wine must be poured into new wineskins [because] no-one after drinking old wine wants the new, for he says "The old is better" ' (Luke 5:38,39). In the past we have found even relatively small adjustments to our accepted (evasive) framework of understanding difficult to make; how much more difficult, then, the adjustment of adopting a whole new <u>unevasive</u> framework of understanding. Thomas Kuhn, a science historian, has said that old scientists who become established within the dominant paradigm (way of explaining/defending/justifying ourself and our world) will virtually never accept the new paradigm, they have to die off before a new idea gains momentum; and Max Planck, a famous physicist, has said 'science progresses funeral by funeral'[1].

Another consequence of our evasive approach in inquiry was that our evasions would become so effective we would lose all awareness that they were evasions — that they served to repress another world and truth. To varying degrees, according to our degree of battle-fatigue/alienation, we would lose memory of that unbelievably beautiful world; that world we deliberately caused ourselves to become 'lost' from in order to one day have it restored permanently to earth. To varying degrees we would forget that we were fighting for the unevasive world. To varying degrees we would come to believe there was no other world but the horrible and unhappy world full of suffering that we were living in. Using our old metaphysical terms again, we would become 'a slave to [the world of] sin' (John 8:34) where 'men loved darkness instead of light' (John 3:19). When such disbelief in another better world, such cynicism, took hold of us we would

---

[1] The Kuhn and Planck quotes were mentioned by Marilyn Ferguson, author of *The Aquarian Conspiracy*, in an interview published in the *New Age* magazine, August 1982.

find ourselves actively resisting, to varying degrees, any recognition or reintroduction of the beautiful world; we would find ourselves advocating staying in the suffering world even though we had initially set out to transcend it!

As well as these 'punch drunk'/embattled/exhausted states there was also the problem of believing it was possible for anyone to dismantle all our evasions. Reflecting or expressing our own inability to confront the truth we would find it hard to see that anyone else could do it. In our evasion of the existence of innocence (because innocence criticised our lack of it), of the less embattled states, we were not aware of what such innocence was capable of seeing and doing. In the introduction to this book it was mentioned that the philosopher Thomas Nagal thought our brains were not made to get to the bottom of the problem of good and evil. When we reflected on our own utterly lost and bewildered state, as Nagal was doing, we could easily become incredulous and pessimistic. We could lose hope and faith that freedom for humanity was possible.

Further, we could become so embattled, our soul could become so repressed, we would lose all ability to recognise what was sound and what was not. In such a state we could become what we call paranoid, unable to know or trust the real truth of a situation. The more insecure we became the harder it was for us to recognise what was sound explanation and what was not. If the mongol hordes of Genghis Khan had been attacking our country for thousands of years and then one day returned to Mongolia it would take many months before we would be able to trust they were actually gone. It would be a brave man who would first venture out of the woods and caves where we had been hiding.

For all these reasons it is expected that initially great resistance and scepticism will meet this book. The very last thing we will believe is that it is what it claims to be — the full liberating truth about ourselves. We will suspect it to be an expression of some form of disguised psychosis and will see its authority, its sense of conviction, as offensive arrogance. (We are not at all familiar with the holistic/subjective/introspective approach. Living with the truth instead of evading it, holistic inquiry can know when it is right and when it is wrong. It is the difference between finding your way around a room with the light on and trying to find your way around it in the dark. Objective inquiry had to grope, as it were, its way round the room step by proven step. Objective mechanistic inquiry was a blind form of inquiry that naturally lacked confidence and authority; we, being familiar only with the mechanistic approach, imagined everyone was similarly insecure.) But this book will survive our disbelief. Normally when we read a book we soon discover its particular level of superficiality. This book however will be different. We will continually think we have found a weakness — that we have been disappointed — that we have 'hit bottom' — only to discover later it was not so. We will not discover its level of insecurity/superficiality/alienation because it doesn't have one, it goes all the way to the full truth about ourselves. That this happens will bear witness to the soundness of the information. 'By their fruits we can recognise them' (Math 7:16).

The Need for Explanation and Adjustment Time

So, although this freedom that the full truth delivers is what we have longed for throughout the ascent of humanity, our first reaction will be to procrastinate. We have been building the staircase for so long we have almost come to believe there is no other life, we have forgotten what the staircase was to achieve, and in some ways we have become so exhausted from the labour we almost feel we don't want our freedom. We have become cynical and embattled. We also have to overcome our fear of being exposed. Further we are incredulous; even though the gate through the wall is opened we can't easily believe it. Nevertheless, while it will take time, we will gradually realise that we have at last succeeded in liberating ourselves and will take our well-earned freedom.

Each of us will need time to adjust. It has to be stressed again that some initial confusion is to be expected in making this adjustment. Earlier in the text when our rehabilitation stage was being described it was mentioned that 'until humanity has mastered the skills that will be necessary, we should limit our efforts to self-expose and to psychologically analyse others to what comes naturally to us as our minds absorb understanding'. To illustrate this problem. In the past it has taken each of us many years to achieve a full recognition of the depth of the pain associated with the human condition. While our basic alienation/personality was established in our infancy and early childhood we did not realise the full extent and consequences of it until we reached the middle and latter half of our life. As an illustration of this, young people often did not appreciate their parents' need to go to Church. Because they have yet to come to this full realisation and thus don't fully recognise the dangers of exposing that alienation, young people naively will tend to want to confront older people — such as their university lecturers and parents — with these explanations, unaware of how deeply evaded these truths are, and have had to be, in us. There is a need for consideration and patience. Until the counselling skills to accompany these understandings are developed, those who are attempting to understand them and their implications should not try to force the unevasive truth upon others. We can give others access to the unevaded truth but they should not have it imposed upon them. We need time to adjust.

When these explanations are first read, the reader (especially those more naive or unaware of the hurtful implications the mention of such truths as integrative meaning have traditionally inflicted upon us) should find them extremely interesting because they account for so many mysteries. Actually they should be so accountable, they should make so much sense, the reader will feel that what is being said is exactly what he or she has known and been trying to say all along. (Huxley's exclamation that Darwin's idea of natural selection should have been obvious to him [Huxley] was an expression of just this kind of recognition.) However, after some months of living with the explanations each person will discover that he or she is being brought back into contact with the deep hurts and evasions that all humans except exceptional innocents have hidden within them. It is these same hurts that left us unable to grapple with, find and reveal these explanations for ourselves. Although it did not start out this way two million years ago, at the end of humanity's search it was alienation not insufficient cleverness that stood between us and the explanation/defence for ourselves. The balance had tipped the other way and we had become too clever, too unsound. Rediscovering and re-negotiating our deeply repressed upsets/alienations/block-outs will be traumatic. As soon as possible video counselling films have to be made showing other people being guided safely through the initial shocking self-confrontation stage. Unless these are available we can all too easily slightly misunderstand — become slightly lost again — and again start seeing ourselves as worthless which will lead us into depression. If we feel worthless we have necessarily misunderstood because the full truth is we are in no way bad/worthless.

One dangerous trap is to fail to realise that these upsets exist in everyone, is to think the repressed upsets that we will discover in ourselves are unique to, or exceptional in us personally. The reason this pitfall exists is that until now humanity has evaded acknowledging the extent of the existence of upsets. It was only a few years ago that going to a psychiatrist became socially acceptable. Unable to tackle our neuroses we have pretended they did not exist. Many of us may not be aware that all humans have similar repressions so we can be left feeling terribly alone with our alienation. Ideally revelation/'exposure' is not to be negotiated alone. Hopefully in the near future institutions manned by staff fully familiar with all the explanations we will need to cope with exposure day will be established for people to attend while they are reading this book. Initially the reality is it will be difficult to find people to look at all the truths that we have practised evading and thus read the book, let alone have them absorb and adjust to the information sufficiently to respond and support what it is about. (Of course eventually humans will be able to grow up familiar with the truth. Instead of schools teaching information that is so evasive it is almost meaningless, as currently occurs, children will be quickly introduced to the predicament of humans on earth so they will not have to cope by adopting evasion. Able to maintain their true selves they will be full of enthusiasm for and love of our world and will learn all that has to be learnt out of interest/fascination with their world and themselves. They will not have to be <u>taught</u> anything.) If we start becoming seriously depressed after reading this book and attempting reconciliation on our own, we need help because we are actually not worthless at all and have either not properly and fully understood the explanation or have been destabilised and disorientated by the amount of information to which we are trying to adjust.

The extreme fragility of our sense of worth or self-esteem, together with the fact that we have evaded the existence of and are therefore unaware of this extreme insecurity within us, are the problems we face in the immediate future. They are the problems associated with 'future shock' mentioned earlier. To quickly give some indication of just how extreme our insecurity/fragility is, consider the following:

For two million years humans have been unable to defend their search for understanding. We have each 'known' our life was totally legitimate and worthy and have been preoccupied trying to establish that fact. We were constantly looking for support and sympathy for ourselves. Real sympathy has not been possible until now so the best we could hope for was to find other people like ourselves and a realm that didn't expose us — we sought people and environments we could <u>identify</u> with. This desperate search for sympathy was often revealed in our choice of mate. How often have couples pictured in engagement and

Photo of W. Watmough which appears at the beginning of his book, *The Cult of the Budgerigar*, 4th edition 1954. This book is regarded as the authoritative work on the breeding of this popular parrot pet (pictured on the right). The author's *affinity* with his *pets* is reflected in his apparently quite unconscious adoption of their side-on way of looking, their round eyes, their tucked chin and their puffed chests. He even *shares* a budgerigar's shiny beak-like-nose and domed forehead.

The caption under this sequence of photos which appeared in the Australian *Womens Weekly* magazine, June 30, 1982, says: 'After 35 years of marriage, it seems that the Queen and Prince Philip have reached that stage where they mimic each other's movement almost instinctively.'

Engagement photo of singer Olivia Newton-John and actor-dancer Matt Lattanzi shows they share identical smiles. (This photo appeared in the Sydney newspaper, *Daily Mirror*, November 16, 1984.)

Heart transplant pioneer Dr. Christian Barnard exhibiting the same distinctive smile as his girlfriend Karen Storay Dowdle. (This photo appeared in the Sydney newspaper, *Sunday Telegraph*, February 20, 1983.)

Pre-wedding photo of Prince Andrew and the then Sarah Ferguson reveals they have similar smiles. (The photo appeared in the *Sydney Morning Herald* newspaper, July 23, 1986.)

wedding photos in newspapers and magazines had the same looks. Their smiles especially were often almost identical. We found support/sympathy for our identity in others who were like us. This desperate need for support was also what made couples who had spent years together adopt each other's mannerisms. It is also the reason we could either take on the identity of our pets or sympathise with and thus select pets that looked like us. Though we cannot begin to imagine how insecure we have become after two million years of being indefensible these illustrations give us a glimpse of just how desperate we were for sympathy.

They also reveal something of the true depth of the difficulty of maintaining our self-esteem or sense of worth during life in alienation. The false defences we built up around ourselves were like a castle built of cards — the structure could all too easily come tumbling down. We 'kept it all together' — but only just — by employing every little bit of support for our egos that we could find and by blocking out any criticism. Now our blocking out is exposed and massive disruptive change is being introduced.

There is a lot to adjust to. While we have at last found the real defence for ourselves (and it is anything but demoralising, in fact it is the ultimate reinforcement for self), the rebuilding process involved is shattering and as such is severely disruptive of our fragile state of security. To protect our fragile stability we will be tempted to reject the information out of hand.

If we are aware of this reaction, this temptation to stay with the old defences, we will be in a strong position to contain it and in so doing will help the process of repair and rehabilitation on and of the earth a great deal. While this early rejection stage will pass as a matter of course, the longer it lasts the longer suffering persists on earth.

Apart from the destabilising effect — which is overcome with sufficient adjustment time — it is not too difficult to get through the initial confrontation stage as long as we are helped with sufficient explanation to allow us to understand. Until now only exceptional prophets could reach an integration-based unevasive and thus full understanding of our upset condition and they couldn't communicate their understanding because humanity had not found the mechanisms to make an understandable explanation possible. The situation now that this full first principle-based understanding is found is we can all share in it and thus eliminate our upset. However the problem is that while this understanding is now available it is not easy for us to use it to understand ourselves. If we were practised holistic thinkers we would only need the outline of the explanation sketched in this book to work out the remaining details for ourselves and apply them to our own particular situation. Not being holistic thinkers and instead ardent evaders of holism we will all too easily become lost when we try to think holistically. To learn to think this way we need lots of help. In fact we need everything explained because being crooked thinkers we will not be able to take the explanations on for more than a few steps on our own before unwittingly reverting to our old evasive explanations and becoming lost again. We are trained to think crookedly not to think straight. We are mental cripples when it comes to thinking holistically. While there is sufficient explanation to adjust to for the present in this book ultimately we will need a great deal more explanation before we can find security of understanding about ourselves in the presence of the truth.

Further, evasion is gradually going to become a useless form of defence. Hiding, escape, self-distraction, block-out, denial, pretence, facades, distracting humour and a bigger car or a new pair of blue shoes or another squillion bucks takeover deal will all eventually become ineffective in relieving the pain in our brain. Confrontation with the pain rather than escape is the solution and for that we need explanation/understanding. 'To live well' will soon no longer be 'the best form of revenge'/defence, rather to understand will be.

For these reasons it is obvious we need to find our prophets — our straight/unevasive thinkers — and bring them to the forefront. They can 'explain everything to us' (John 4:25) and explanation is all important now. The 'last [the innocent] need to be first' (see Math 19:30; 20:16; Mark 10:31: Luke 13:30) during rehabilitation. Egotism — the need to defend ourself against criticism — is finished with. There is no longer any criticism of anyone. Demonstrations of our worth are not necessary now. We can understand that we are worthy. We can love ourselves and others now. Those most capable of explaining can best show us how to do this because doing it depends on explanation/understanding.

It is hoped that many people will come forward to assist this project, if not straight after they have read this condensed version then once it becomes obvious that what is being foreseen here is actually beginning to happen. Then proper preparation can be made for the traumas that revelation/confrontation/exposure/'judgement-day' brings for many people.

Until proper preparations can be made for exposure day — for all-out self-confrontation — it will be necessary for each of us to rely on block-out or evasion to cope because if we can't understand and/or adjust to the understanding and can't confront the truth then evasion, as before, is all that is possible. While humanity is getting itself properly organised to cope with revelation day we each will need to be free to hide from the truth as much as we need to. (This will happen anyway.)

After negotiating the initial shock of self-confrontation we will each back off from the truth again, repressing it to varying degrees, re-establishing our self at a distance from the truth at which we feel comfortable. From this new position we will gradually begin to think ourselves together at a rate that suits us. If someone asks us about the truth we will say 'I know about as much as I want to know for the present but you are welcome to read it for yourself. The book is over there on that shelf. Take it, come to your own terms with the material then come and play a round of golf (or whatever) with me — but leave me to adjust to the information at my own pace'.

It will take us time to heal and each of us will be starting at a different distance from the truth. The less battle-weary, being less distanced from the truth, will have less ground to travel but the more exhausted, since they tend to be the more clever among us, will be able to think faster and absorb the understandings more quickly so they should be able to heal themselves faster, the net result being that we should all arrive at soundness at approximately the same time. Ultimately the meek [the innocent] won't inherit the earth (see Math 5:5) although initially during rehabilitation it is true that they will. During rehabilitation the innocent will lead the battle-weary, will show the way — the 'last will be first'.

It might be emphasised again here that in the final analysis differences between people and races in the degree of their battle exhaustions will be insignificant. We are entering a world of so much beauty and happiness all remaining problems will be drowned out. In fact they will not be a problem to us, rather they will be a pleasure to overcome. Humanity will have so much generosity nothing will be a problem. In our exhausted cynicism we have often tried to turn this positive into a negative by saying that 'the price of paradise will be boredom' and that 'there is only fun in the adventure/journey itself'. (For example, from an essay on secrecy which appeared in *Time* magazine, January 17, 1983, 'Still, anybody contemplating humanity as it is must wonder whether, in a thoroughly transparent world, the species would not suffer spiritual anemia and perhaps terminal boredom'.) As was explained on page 43 we have to remember that the excitement of adventure was 'in truth a very small positive'. We have lost virtually all memory of the real 'positives' our world has to offer. Our cynicism about the future was a result of loss of access to what true happiness is like and in the extreme a case of painting the future black to make ourselves feel better about our own bleak state. Like many of our negative attitudes, it was a false form of self-defence and one of the products of extreme exhaustion.

The rapid and major adjustment phase of our journey home to soundness will take possibly two or three generations with the finer adjustment taking a few generations longer. For our own immediate future, while it will take time to fully confront, acknowledge and adjust to the truth, we will very soon derive great relief from the realisation that we at last have the full/compassionate truth up on earth. Our hope and faith that humanity would achieve liberation and avert apocalyptic self-destruction has been realised. With this realisation we can find the strength/enthusiasm to call a moratorium on our upsets and buy the time necessary to heal ourselves and our planet. Our great anxiety for ourselves, our children and our world can at last subside. We have made it to safety. Everything is possible now. Essentially, while we still have to repair ourselves and our earth, we have achieved our freedom.

The Centre For Humanity's Adulthood

Of course, while the truth is available now on earth in this book we are not safe until all humanity has been made aware of it.

The Centre For Humanity's Adulthood has been formed to help communicate and develop these understandings. Now that the door to humanity's adulthood has been opened there is suddenly an enormous amount of work that can and must be done. We are no longer powerless to solve the suffering on earth. We now have the means to solve upset and end the need for weapons. At the time of printing this book the C.F.H.A. has a tiny staff and very limited funds. We hope people will offer support so that we can grow and grow quickly.

Science has done its job of finding all the pieces of the jigsaw of explanation and religions have done their part in maintaining us but neither of these institutions could truthfully assemble the jigsaw, deliver it to the world and set about clearing up all the upset, anger and heartache it took to get here. Science was an evasion and religion a retreat; now we want confrontation for our rehabilitation. While traditional science and religion are huge institutions with tremendous resources, the art and means of rehabilitation — of confrontation with our exhaustions — has yet to have a single brick laid for it. There is no centre or institution for unevasive science to which this work can be taken for support, recognition or endorsement. For instance, there are no journals for unevasive thinking which can acknowledge, present and promote this work. Consequently it has had to be independently funded, published and promoted. It is wrong to ask for help (in an ideal world help is automatically forthcoming) but in our exhaustion if things are not said we often do not recognise them.

Because of the insecurity that exists amongst humans many will strenuously deny the ideas in this book. In the past the way we attacked innocent thought was firstly to deny its insights then, if that failed, try to discredit its source. Finally, if these devices failed we did nothing — made no comment and gave no response and in so doing let the information die. While the information in this book is innocent thought it has reached all the way to the full truth — it does not criticise us as innocent thought almost always has done in the past. Nevertheless, even though it is safe we will still tend to treat it the way we have always treated innocent thought. We are professional evaders now and rightly incredulous. (To quote Anthony Barnett, Professor of Biology at the Australian National University: 'In all written history there are only two or three people who have ever been able to think on this [the macro] scale about the human condition.'[1]) The storm of this denial stage has to be ridden out. Earth and its people badly need relief from our embattled state and this book provides it.

Our insecurity has to be resisted in other ways as well. Because of our insecurity we have preferred that no one 'rock the boat' — that all that is said in this book not be said — that no one be made to feel awkward and uncomfortable. But what is said in

---

[1] In conversation with the author in Canberra in January 1983. After saying that only two or three people have ever been able to grapple with the Human Condition, the professor added that therefore he 'wasn't about to believe that [this author was] another of them.'

this book had to be said. Someone had to stand up perfectly straight, walk right down the middle, resolve the upset, set us free (overturn our evasions, break the silence, let the light in upon our preferred darkness) and stop the terrible suffering of people and wildlife and the earth itself that is going on everywhere almost unacknowledged. (Another way of saying this is that someone had to go off and sit down quietly under the Banyan/thinking tree for a few years, carefully think it all out from the beginning and then come back and tell us the answers that we have come to need but have not had the time or inclination to think about.)

This book has to be circulated, promoted and, if they can be interested and we can agree, its Summary printed in widely read influential journals. Distribution centres have to be established in the U.S.A. and in England. Then the full book has to be published, translations undertaken, video explaining and training films made. Eventually we want to establish Centres For Humanity's Adulthood somewhere in the wilderness of each continent — in Australia, America, the Rift Valley of Africa, Asia and Europe. The way forward is now back — back to our soul's world of nature to heal our hurt selves. We can make peace with our soul now and nature is its closest friend. Nature will look after and heal us as we come home from the war. Our soul, that child within us, is going to be so excited to see all its friends again. It will lovingly take our bruised and battered hand and show us round its home.

Humans have to stop turning outwards, escaping, self-distracting, adventuring, being entertained and in general being superficial, and start turning inwards and confronting/addressing the real problem on earth of our upset selves. This was not possible in the past but it is now, since we have the understanding necessary to do it. Rehabilitation of ourselves and our earth can begin.

You can help get the project underway by giving a copy of this book to a friend or someone you think might be interested or even sending us the address of someone influential who we should send a complimentary copy to. Copies are available through bookshops or the book's distributor in Australia *Bookchain Australia Pty Ltd*, Box 213, Brookvale, N.S.W. 2100 Tel: Sydney (02) 9385455 or Fax (02) 9385904, or you can write to the *Centre For Humanity's Adulthood* at the address on the next page where the books are available at the recommended retail price in 1988 of $12 plus $2 postage each (for overseas purchases postage should be calculated on the weight of the book plus some extra weight for wrapping).

If you would like to support the C.F.H.A. you are invited to become a member of the C.F.H.A. Membership fees at the time of publishing are $25 a year and you may enrol for as many years as you wish. For $500 you receive Life Membership. All enrolments will be acknowledged and members will receive copies of the C.F.H.A. newsletter when it comes out. We hope the newsletter will quickly grow into a magazine that answers readers' questions. Members will also be informed of the publication and price of the full book when it becomes available and of the availability of video explaining films as well as other facilities. Possibly a super cheap version of this book could be printed on cheap paper and stapled like a comic. These could be distributed at universities and schools. As soon as possible we hope to be in a position to help those interested establish *Workshops For Humanity's Adulthood* where people can go to meet and talk with others about the development of their adulthood understandings.

The extent of the support the Centre receives from membership fees and donations will dictate how quickly the infrastructure for the Centre can be established.

A registration form for membership is included on the next page.

Registration form for Membership in the
Centre For Humanity's Adulthood

- I include payment of $ _____ for _____ year(s) membership (at $25 per year or $500 for life membership.)

- Please send me _____ copies of *Free: The End of the Human Condition* at the recommended retail price of $12 plus $2 postage per book within Australia. (Note that this is the price in 1988 only. Beyond 1988 you might estimate the price change as $12 is the typical price for a book of this size in the bookshops in 1988 or you can contact the Centre for the updated prices.)

I am paying by ☐ cheque money order for $ _____

The Centre for Humanity's Adulthood (CFHA) became a registered charity under the name Foundation for Humanity's Adulthood (FHA) in 1991.

Card No.

With the advent of the internet we no longer publish newsletters and no longer have subscription as developments and information now appear on our website

Signature **www.humancondition.info**

Name **Telephone** +61 2 9486 3308, **fax** +61 2 9486 3409.
email info@humancondition.info or **write** to
GPO Box 5095, Sydney NSW 2001, AUSTRALIA.
(The box number ensures the Foundation a permanent address should its location and phone number change.)

Send to: Centre for Humanity's Adulthood
GPO Box 5095
Sydney NSW 2001
Australia

(We regret the impersonal box number however the Centre needs an address that will stay operational wherever the Centre is located.)

- To avoid tearing out this page we suggest you write the details out separately or preferably photocopy the form.

Index